A Correlation of Ordovician rocks in the British Isles

ALWYN WILLIAMS, ISLES STRACHAN,
DOUGLAS ANTHONY BASSETT,
*WILLIAM THORNTON DEAN, JOHN KEITH INGHAM,
ANTHONY DAVID WRIGHT &
HARRY BLACKMORE WHITTINGTON

CONTENTS

SUMMARY

A general introduction outlines the problems of defining the Ordovician systemic boundaries, determining the ranges of the type British Series, and correlating the standard British succession with those of Bohemia, Estonia, Kazakhstan, North America and Australia, the last of which is also considered separately in the context of graptolite zonation. These sections are followed by charts with explanatory statements showing the correlation of selected Ordovician successions in Wales, the Welsh Borderland, eastern and northern England, Scotland and Ireland. All contributors have adopted an agreed time-stratigraphic framework for the charts; but although there is general concurrence on the main conclusions insofar as they affect intra-British correlation, detailed correlation within each region is the responsibility of the named contributor(s).

---

1. Introduction and general aspects of correlation

By ALWYN WILLIAMS

ALTHOUGH British rock successions were used by Lapworth in 1879 to found the Ordovician System, they have never been presented in a comprehensive correlation that has won general and lasting approbation. Even attempts to do so have been comparatively rare, those of Watts (*in* Evans & Stubblefield 1929) and Whittard (1960) being two notable exceptions. But inconsistencies in the former's use of time-stratigraphic units and the inability of the latter to classify Scottish rocks precisely according to the standard Series of the Ordovician, identify the two main factors militating against the provision of a generally acceptable correlation chart. They are: various persistent controversies concerning the stratigraphic delineation of systemic and infra-systemic boundaries, and a diversity of biofacies and lithofacies that is surely unsurpassed for so small an area as the British Isles.

---

*Who publishes by permission of the Director, Geological Survey of Canada.

(A) SYSTEMIC BOUNDARIES

The events leading to the location of the Cambrian-Ordovician boundary at either the base of the Arenig as favoured mainly in Britain, or the base of the Tremadoc as practised elsewhere, have already been discussed by Whittington & Williams (1964) and need only a brief review here. When Lapworth (1879a, pp. 13–14) proposed the Ordovician System, he unambiguously cited a type section in the Arenig-Bala district, Merioneth, composed of about 3660 metres of volcanic and sedimentary rocks (*see* Bassett herein). Both the lower and upper boundaries of the succession are well defined (Fig. 1). The former coincides with the Basal Grit of Sedgwick's 'Arenig Group' which rests unconformably on the Tremadocian Amnodd Shales (Fearnsides 1905). The latter is represented by the top of the Foel y Ddinas Mudstones, the youngest Formation of Sedgwick's 'Bala Group' which is succeeded by the Llandoverian Cwm yr Aethnen Mudstones (Bassett *et al.* 1966). However in describing the graptolitic faunas characterising his newly founded System, Lapworth (1879b) incorporated a time-stratigraphic error perpetrated by Hicks (1875) during his studies of the Cambrian-Ordovician rocks of St. David's, Pembrokeshire which was to have far-reaching repercussions. In 1872 Hicks collected a didymograptid graptolite assemblage from beds at St. David's which he assigned to the 'Lower Arenig' but *correlated* with the Upper Tremadoc of North Wales. There were no grounds whatsoever for this stratigraphic folly especially since the type sections of both Series are in North Wales where the 'Arenig Group' had already been shown by Sedgwick (1852) to be younger than the 'Tremadoc Group'. Indeed, it has since been demonstrated (Cox *et al.* 1930, p. 259) that the so-called 'Lower Arenig' or 'Upper Tremadoc' rocks of St. David's are actually correlatives of other local successions identified as 'Middle Arenig' by Hicks, and that no beds of Tremadocian age occur in the district. Nevertheless, when Lapworth (with Hopkinson 1875) described these 'Lower Arenig' graptolites he did not challenge Hicks' assertion that the assemblage was 'Upper Tremadoc' in age. On the contrary, having accepted Linnarsson's correlation (1876, p. 149) of the *Ceratopyge* Limestone of Sweden with the Upper Tremadoc of North Wales, Lapworth (1879b, p. 455) was willing to include that Formation within the 'Lower Arenig' of Hicks and, therefore, within the Ordovician System. In fact, by the beginning of the 20th century, as his discussion of Groom's paper (1902, p. 148) on the English Cambrian showed, Lapworth was more concerned with how much of the Tremadoc Series should be included in the Ordovician than with the precise range of the Arenig Series that he had initially proposed as the base of the System.

Since then many attempts have been made (*see* Whittington & Williams 1964) to muster palaeontologic and stratigraphic evidence favouring the inclusion of the Tremadoc Series in the Ordovician. But as is to be expected in any objective appraisal of all relevant data, the base of the Tremadoc is not better suited to serve as the Cambrian-Ordovician boundary than the base of the Arenig. In such circumstances the historical priority of Lapworth's original proposal for a new system based on a clearly defined type section, which is valid according to any rational Code of Stratigraphic Nomenclature, must be weighed against the greater popularity currently enjoyed by another version of the System amended by an inadvertent perpetuation of a stratigraphic error. Accordingly, until this difference is resolved by international agreement, the Lower Palaeozoic Era Subcommittee of the Geological Society of London provisionally accepts the base of the Arenig as the base of the Ordovician System (*see* Holland 1971).

The stratigraphic location of the Ordovician-Silurian boundary has never been seriously disputed (Whittington & Williams 1964, p. 247). Yet it cannot be confirmed by direct correlation between relevant successions in the Bala and Llandovery districts, and the biostratigraphy of Kazakhstan sections suggests that there are discrepancies in the ranges of diagnostic shelly and graptolitic assemblages.

Exposures of the Foel y Ddinas Mudstones near Bala, with their distinctive *Hirnantia* fauna, constitute the type section for the terminal Ordovician stage, the Hirnantian. They are succeeded by the Cwm yr Aethnen Shales with *Glyptogr. persculptus* (Pugh 1929, p. 274) indicative of the basal graptolitic Zone of the Silurian. In the type area for the Llandovery Series, however, both the topmost Ordovician and basal Silurian beds are badly exposed successions of shelly facies lacking diagnostic faunal assemblages. Cocks *et al.* (1970, p. 81; 1971, p. 104) have, therefore, proposed that the continuous succession of graptolitic shales exposed at Dobb's Linn, near Moffatt be taken as the type section across the Ordovician-Silurian boundary. There, the presence of the *Glyptogr. persculptus* Zone only indicates that some part of the Barren Mudstones Group with 'a few thin graptolitic bands' (Toghill 1970b, p. 4) may be reasonably correlated with the Foel y Ddinas Mudstones. In any event this device is an inadequate solution to the problem because the relationship between the Hirnantian and early Llandoverian shelly faunas is still unknown. Consequently, sequences in north-east or south-west Wales may ultimately prove more helpful, especially the Garth area where, as recently re-iterated by Skevington (1969), graptolites indicative of the *persculptus* Zone occur below the local lithologic base of the Llandovery and are contemporaneous with brachiopods and trilobites which, according to Andrew's identifications (1925, p. 391), are more Ashgillian than Llandoverian in affinities. A similar mixture of an Ashgillian *Dalmanitina mucronata* shelly assemblage and Llandoverian *persculptus* Zone graptolites is reported from Kazakhstan (Nikitin 1971, p. 341). In view of these anomalies, some revision of the correlation between the graptolitic and shelly faunas straddling the Ordovician-Silurian boundary might prove necessary.

(B) CORRELATION OF BRITISH ORDOVICIAN SUCCESSIONS

With the Tremadoc Series at least temporarily assigned to the Cambrian, the Ordovician is divisible into five Series. When Lapworth first erected the System, he referred only to the two divisions of Arenig and Bala. But this restriction was probably out of deference to Sedgwick's researches in North Wales because Murchison's 'Llandeilo' and 'Caradoc' Groups were soon employed by him (1879b, p. 455) as major Ordovician divisions and he also encouraged Hicks (1881) to propose 'Llanvirn' for a group of rocks previously assigned to the Llandeilo and Arenig. Indeed the use of the term 'Bala' as a time-stratigraphic division has never been widely accepted. By 1905, Marr (p. lxxxv) had introduced the Ashgill

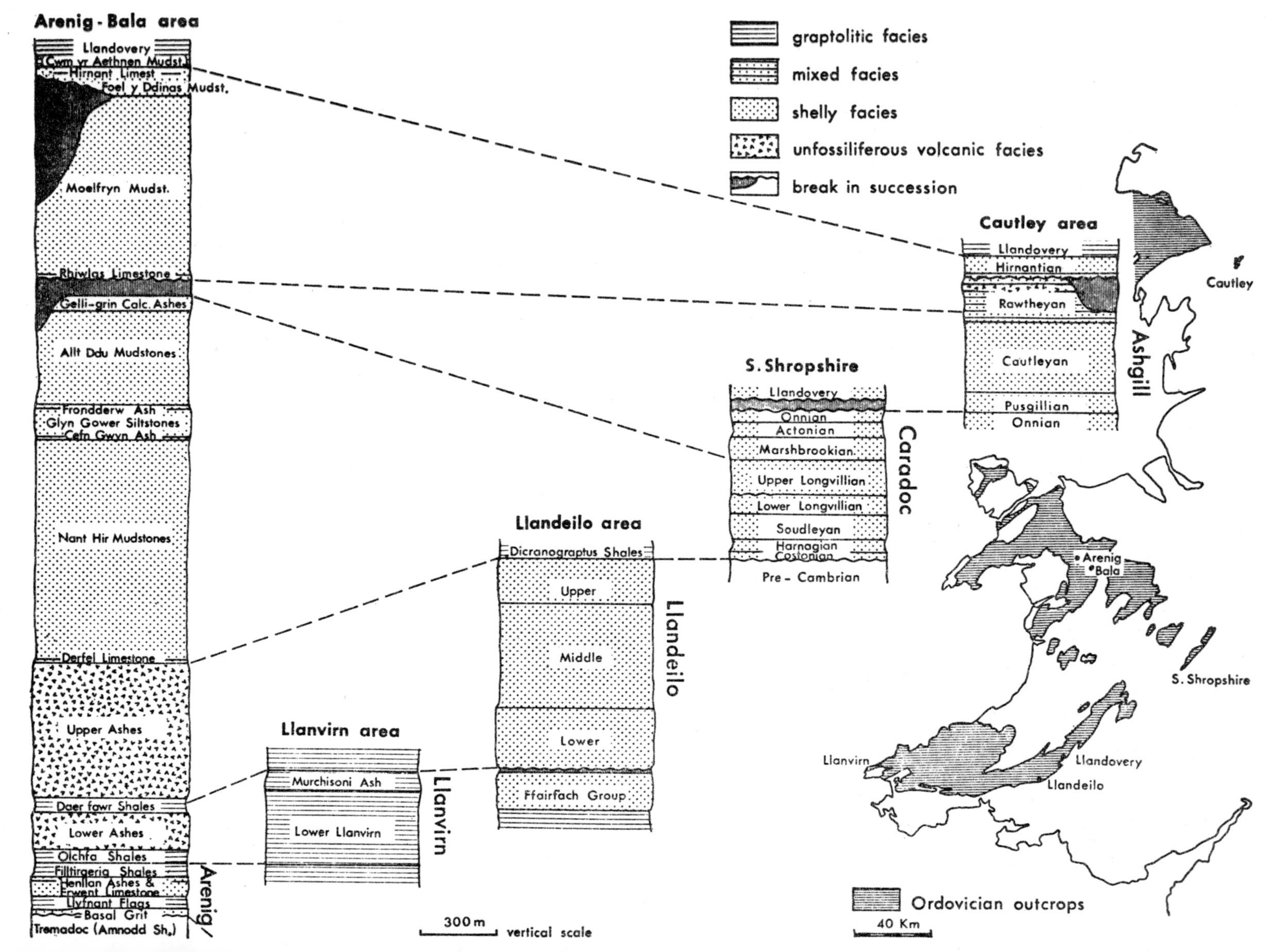
Arenig-Bala area
Llandovery
(Cwm yr Aethnen Mudst.)
Hirnant Limest.
Foel y Ddinas Mudst.
Moelfryn Mudst.
Rhiwlas Limestone
Gelli-grin Calc. Ashes
Allt Ddu Mudstones
Frondderw Ash
Glyn Gower Siltstones
Cefn Gwyn Ash
Nant Hir Mudstones
Derfel Limestone
Upper Ashes
Daer fawr Shales
Lower Ashes
Olchfa Shales
Filltirgeria Shales
Henllan Ashes & Erwent Limestone
Llyfnant Flags
Basal Grit
Arenig
Tremadoc (Amnodd Sh.)
graptolitic facies
mixed facies
shelly facies
unfossiliferous volcanic facies
break in succession
Cautley area
Llandovery
Hirnantian
Rawtheyan
Cautleyan
Pusgillian
Onnian
Ashgill
Cautley
S. Shropshire
Llandovery
Onnian
Actonian
Marshbrookian
Upper Longvillian
Lower Longvillian
Soudleyan
Harnagian
Costonian
Caradoc
Pre-Cambrian
Arenig
Bala
Llandeilo area
Dicranograptus Shales
Upper
Middle
Lower
Ffairfach Group
Llandeilo
Llanvirn area
Murchisoni Ash
Lower Llanvirn
Llanvirn
Llanvirn
Llandovery
Llandeilo
S. Shropshire
300 m vertical scale
Ordovician outcrops
40 Km

Series to embrace the youngest Ordovician successions; and the discovery of a big break within the Bala succession in the type area (Bancroft 1928) prompted Bassett *et al.* (1966, p. 226) to suggest that the use of the 'Bala Series' as a standard infra-systemic division of the Ordovician can no longer be justified.

The relationship between consecutive Series of the Ordovician (Fig. 1) is complicated by the geographic separation of type sections and facies differences. Although the Arenig Series is based on a shelly-graptolitic succession in Merioneth and the Llanvirn on a graptolitic succession in Pembrokeshire, the boundary between the two Series is traceable in the field in both areas as a sharply defined distinction between beds containing the *Didymogr. hirundo* and *bifidus* faunas. The Llanvirn–Llandeilo boundary, on the other hand, is less easily located because of the exclusively shelly nature of the Llandeilo facies in the type area in eastern Carmarthenshire. There the Llandeilo Flags as understood by Murchison (1839) succeed unconformably a diverse suite of fossiliferous rocks (the Ffairfach Group) which pass downwards into shales with graptolites of the *bifidus* Zone. Bergström (1971b, p. 109) has correlated conodont faunas recovered from the top of the Group with those corresponding to the middle part of the equivalent of the *Glyptogr. teretiusculus* Zone in Sweden. But macrofaunal researches by R. Addison on the south Welsh Llandeilo facies and by Whittard (1955-67, p. 306) on the trilobites and the writer on the brachiopods of west Shropshire, suggest that the base of the Meadowtown Beds is coeval with the base of the Llandeilo in the type area. The Meadowtown Beds conformably succeed the Betton Beds (Dean herein) with a mixture of graptolites belonging to the *Didymogr. murchisoni* Zone and shelly assemblages found in the Ffairfach Group and its correlatives in south-west Wales. This correlation upholds the current practice of equating the Ffairfach Group with the *murchisoni* Zone, so that the base of the Llandeilo may be taken as coincident with the base of the *Glyptogr. teretiusculus* Zone, as developed in its type area of south-west Wales.

The relationship between the Caradoc and Llandeilo Series is no longer the controversial issue it was even a few years ago (Williams 1969b, p. 241), thanks largely to studies of relevant British macrofaunal and microfaunal assemblages by R. Addison and S. M. Bergström respectively. In 1953, the writer (p. 194) described the Llandeilo succession in the type area as being succeeded conformably by calcareous shales yielding graptolites attributable to the Zone of *Nemagr. gracilis*. Westwards the Llandeilo facies pass into a mainly graptolitic succession known as the Hendre Shales (Addison *in* Bassett herein). These are succeeded by calcareous shales, the Mydrim Limestone, which have long been known to contain *Nemagr. gracilis* (Hall) and its associates (Evans 1906, p. 631). Jones (1956, p. 327) stated that these graptolites also occurred in the Hendre Shales; but he did not cite any locality and, on the assurance of the Geological Survey (Cantrill & Thomas *in* Strahan *et al.* 1909, p. 46) that both *Didymogr. superstes* Lapworth and *Nemagr. gracilis* were restricted to the Mydrim Limestone, it seemed reasonable (Williams 1969b, p. 243) to correlate the Llandeilo succession of the type area with the Hendre

---

FIG. 1 Biofacies classification and correlation of standard sections of the Ordovician Series of England and Wales (after Williams 1969b; Ingham & Wright 1970).

Shales. This assumption was also made by Toghill (1970a, p. 123) who, in answer to Skevington's proposal (1969, p. 164) to suppress the *Glyptogr. teretiusculus* Zone, pointed out that the lower beds of the Hendre Shales yield a distinctive graptolite fauna that cannot reasonably be accommodated within the *murchisoni* or *gracilis* Zones. Skevington's reservations about the relationship between the Llandeilo Series and the standard graptolitic sequence, and Toghill's demonstration of the individuality of the older Hendre Shales graptolitic faunas, are compatible with the conclusions of Addison in his study, shortly to be published, of the successions west of Llandeilo. Near the type section of the Hendre Shales, Addison has found *Nemagr. gracilis* well down in the succession. Associated shelly faunas, here and in the vicinity, include the Lower Llandeilo *Lloydolithus lloydi* (Salter) and indicate that the base of the *gracilis* Zone should be brought down to coincide, more or less, with the top of the Lower Llandeilo. Moreover, further west in the Narberth area, Addison has demonstrated that shelly faunas characteristic of the type Llandeilo Series are succeeded by assemblages of undoubted Costonian age. But as Dean points out herein, the *gracilis–multidens* boundary is best located within the upper part of the Costonian Stage. Consequently the range of the *gracilis* Zone must now be greatly extended to include beds of middle and late Llandeilo as well as Costonian age, and the range of the *teretiusculus* Zone correspondingly restricted.

The shelly faunas of the Caradoc Series in the type area are reasonably well known and afforded Bancroft (1933, 1945) the means of establishing seven Stages. An extra Stage, the Pusgillian, was later proposed for sections in northern England but is now incorporated in the Ashgill Series. The relationship of the post-Costonian stages with the *Diplogr. multidens* and *Dicranogr. clingani* Zones, however, cannot be precisely defined (Dean herein); while the numerous brachiopod and trilobite zones on which the stages were founded have a restricted application. For example, they cannot be used in the Bala area where only a few species, also found in the standard section of Shropshire, aid correlation (Bassett *et al.* 1966, p. 256). This deficiency in Bancroft's zonal scheme is probably attributable to lithofacies control (Williams 1963, p. 343) although it can be greatly reduced by using assemblage zones to determine stage equivalents or divisions thereof.

With the transfer to the Ashgill Series of the Pusgill Stage, which is not represented in any known Shropshire section, the Caradoc-Ashgill boundary is now widely acknowledged to be satisfactorily located within the British shelly successions. Its position within the standard graptolitic sequence, however, is less certain. Cave (1965, p. 287) concludes that the graptolites associated with *Onnia gracilis* in the Pen-y-Garnedd Shales of Welshpool show the Onnian not to be younger than the *Dicranogr. clingani* Zone. Toghill (*in* Ingham & Wright 1970, p. 235), on the other hand, considers the same assemblage, on balance, to belong to the *Pleurogr. linearis* Zone. This assessment is in keeping with evidence in the Girvan area that the Upper Whitehouse Formation with *Dicellogr. complanatus* (Toghill 1970b, p. 14) is partly Pusgillian in age (Ingham & Wright 1970, p. 234).

The Ashgillian has been divided into four Stages characterised by diagnostic assemblage zones (*see* Ingham & Wright herein). The terminal assemblage indicative of the Hirnantian Stage is particularly distinctive and occurs as an easily recognised fauna in many parts of Eurasia and North America.

(c) CORRELATION WITH ORDOVICIAN SUCCESSIONS OUTSIDE THE BRITISH ISLES

In correlating the standard Ordovician sections of the British Isles with those of other regions we must now recognise that no one faunal succession can serve as a ubiquitous time-stratigraphic measure. Even the familiar sequence of graptolitic zones based on British assemblages are not strictly applicable in Britain itself. In particular, the replacement of the *Climacogr. peltifer* and *Climacogr. wilsoni* Zones in the Southern Uplands by the *Diplogr. multidens* Zone in Shropshire and three other diplograptid (*s. l.*) Zones in south-west Wales has long been advocated and practised (*see* Skevington 1969, p. 166). Benthic faunas, as is to be expected, generally play a much more restricted role in detailed correlation, although broadly based assemblage zones, usually at the generic level, can be recognised over large areas. The presence of the Nevada-Newfoundland Whiterock fauna in western Ireland (Williams 1972a) and of the Appalachian Porterfield fauna in south west Scotland (Williams 1962) are two examples.

On the more speculative side, evidence is now accumulating to suggest that most, if not all, phyla were differentiated into a few very large realms (or provinces), biostratigraphic traces of which have subsequently been fragmented and differently reassembled by recurrent continental drift. Consequently successions of assemblage zones of varying taxonomic precision are simply identifiable phases in the evolution of a relatively few provincial faunas (Williams 1969a; Whittington & Hughes 1972). Such faunas may retain their individuality for a considerable time subject only to inherent changes arising from speciation and extinction affecting constituent taxa. But usually the cumulative affects of recruitment from, and exchange with, constituents of other provinces induce gross changes in the composition and geographic distribution of faunas, and at the same time provide the means of effecting correlation even by use of benthic taxa. The widespread occurrence of the Hirnantian assemblage (Ingham & Wright herein) as a prelude to the seemingly cosmopolitan nature of the Llandovery shelly faunas (Berry & Boucot 1970) is a case in point.

It is, therefore, possible to assemble successions into Series and Stages, mainly based on the provincial characteristics of the fossil faunas, to serve as time-stratigraphic standards for very large areas. Five classic sections have been chosen to represent world Ordovician successions and have been tentatively correlated with the standard British succession. The results are given in Fig. 2. All faunas have been used to effect as accurate a correlation as possible, although graptolites continue to offer the most reliable means of establishing contemporaneity. Recent intensive studies of Ordovician conodont faunas have prompted controversial as well as informative correlations (for example, Bergström 1971b). Some of the correlations are inconsistent with those derived from comparisons of shelly and graptolitic faunas; and although they may ultimately prove to be more exact, it is important to remember that conodontophoride animals, whatever they may have been, must also have been subjected to some ecological control and were, therefore, at least as susceptible to provincialism and diachronic distribution as graptolites. The most reliable correlation is always likely to be a compromise of the relative age determinations of all phyla represented within a succession. Consequently, some of the more disputable correlations advocated by conodont researchers and identified below have not been used in Fig. 2 pending further researches in all assemblages.

The biostratigraphy of the Bohemian Ordovician succession is well known especially through the researches of Havlíček, and his review (with Vaněk 1966) forms the basis of the correlation given here. The Bohemian shelly assemblages show some similarities with those of southern Britain and, to a lesser degree, the Baltic. But it is essentially a highly distinctive fauna probably extensively distributed throughout the Mediterranean countries, the Middle East and North Africa. Despite differences in species assemblages, the graptolites of the Sárka Formation can be confidently correlated with the Llanvirn and those of the Loděnice and Bohdalec substages with the *wilsoni* to *clingani* Zones inclusive. Also, with the extension of the *gracilis* Zone to include part of the Llandeilo Series, ambiguities in the correlation of the Dobrotivá Formation and Chrustenice Substage (Havlíček & Vaněk *op. cit.* p. 37) have now been resolved. The precise location of the Caradoc-Ashgill boundary within the Bohemian section, however, has yet to be determined. *Dicranogr. clingani* Carruthers is known in the Bohdalec Formation while *Dicellogr. anceps* Nicholson occurs throughout the Králův Dvůr Formation which also contains a diagnostic Ashgillian shelly fauna. Yet no representatives of the graptolitic *linearis-complanatus* or shelly Pusgillian faunas have been found, and correlation of this part of the British section with the upper part of the Bohdalec Formation is open to revision (Dean 1967, p. 319). In contrast, the Kosov Formation contains the Hirnantian fauna and permits an accurate determination of the Ordovician-Silurian boundary.

The Ordovician rocks of Estonia constitute another classic section which serves as a standard for the peri-Baltic Shield facies with representatives of its fossil assemblages occurring sporadically in North Wales and eastern Ireland. The rich fauna is dominantly shelly and is the biostratigraphic foundation for three Series comprising sixteen Stages. Intensive studies of faunas and sediments have led to continual refinement in the definition of the Stages, although correlation with the British section has been fairly consistent. The section given in Fig. 2 incorporates the researches of Männil (1966, 1971) and Rõõmusoks (1960, 1970) and is based mainly on the ranges of conodont, chitinozoan and macroscopic shelly assemblages. In the light of these data, the Oeland Series terminates at the top of the *bifidus* Zone, while the Lasnamägi Stage appears to straddle the Llanvirn-Llandeilo boundary. The Kukruse Stage is satisfactorily defined as coincident with the range of *gracilis;* and recent chitinozoan studies (Männil 1971, p. 310) confirm a Soudleyan age for at least part of the Jõhvi Stage. In younger successions, the only correlation that can be advocated with some confidence, according to Männil (*op. cit.*, p. 311), is the Caradoc-Ashgill boundary with the upper part of the Vormsi Stage.

Although the Ordovician successions of Kazakhstan are currently less well known than other sections drawn in Fig. 2, they are destined to play an important role in international correlation. They contain graptolitic and shelly assemblages bearing strong affinities with those of Britain on the one hand and of the Siberian Platform and western North America on the other (Nikitin & Apollonov 1968). Consequently the correlations proposed by Apollonov *et al.* (1968) and the

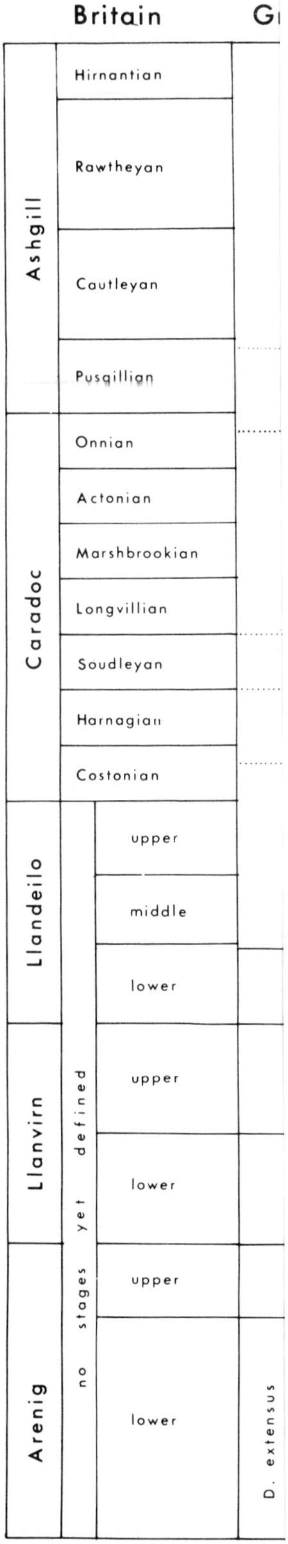

FIG. 2—Chart showing a c
classic sections in other

shelly faunas of later Canadian and early Champlainian age (Berry 1967, 1968; Skevington 1968). However, the balance struck by Whittington (1962-68, pp. 93-138) after considering both shelly and graptolitic assemblages seems to be most reasonable and his correlation has been used in compiling the older part of the North American standard section.

Notwithstanding the complications outlined above, some key inter-continental correlations are acceptable to most biostratigraphers. The Whiterock–Marmor Stages probably overlap with the Llanvirn; the Porterfield is likely to be equivalent to the *gracilis* Zone, and the Eden–Maysville junction to the Caradoc–Ashgill boundary. Further researches, especially with the prospect of being able to integrate correlations provincially, should provide other reliable age determinations.

(D) LAYOUT OF CORRELATION CHARTS

A left-hand column of standard Series is common to all the charts. The Lower Ordovician Series are subdivided informally in all but the Scottish chart, the accepted Upper Ordovician stages are shown on all charts. The ranges of the standard graptolite zones are indicated in Figs. 2 and 3.

Correlation of conformable boundaries of formations or other stratigraphical units are shown in descending degrees of certainty by solid, broken and dotted lines, a question mark indicating the greatest uncertainty. An unconformity at the base of a formation is represented by a wavy solid line. Vertical lines indicate gaps in succession and their length shows the stratigraphic extent of the interval. Blank spaces imply that strata of this age are unknown but may be present. A solid line running vertically or diagonally indicates an erosional surface; surfaces of uncertain nature are shown by broken or dotted lines. Zig-zag lines are used to denote either boundaries between facies or an unconformity with overstep and/or overlap.

New formational names have not been introduced for rock-stratigraphic units, but the term 'Series' has been replaced by Group and, in many cases, 'Beds' by Formation. This follows the practice of the American Commission on Stratigraphic Nomenclature (1961).

2. Correlation of British and Australian graptolite zones

By ISLES STRACHAN

THE general sequence of Ordovician graptolite zones in Britain and in Australia has been established for a long time. In the period 1850–1900 most of the forms recorded from Australia were matched with species recorded first from North America or Europe and the world-wide range of graptolites accepted. With the advent of a more critical approach to species (e.g. Hall 1899, p. 440) which resulted from various revisions in Europe, North America and Australia, differences between the areas were emphasised although rarely with adequate comparison of type material. Later, Monsen (1937) and Ekström (1937) both recorded forms allied to Australian species from Scandinavia and in recent works more and more of these have been recognised in Europe so that comparison of the faunas is again becoming possible. In his recent review of graptolite faunal distribution, Bulman (1971) has pointed out the cosmopolitan nature of some species and the restricted appearance of others. It is only the widely distributed forms which can be used for long distance correlation but even they must have had a local origin and their time-distribution be diachronous. Precise local correlations may indicate centres of origin but such detail is not yet available while so many of the first described species are, by modern standards, inadequately known.

There has been considerable discussion recently on the correlation of European and North American Ordovician graptolite zones (Berry 1968; Skevington 1968) with apparently irreconcilable conclusions being reached. The problem of matching British and Australian faunas has therefore been tackled directly to see if there are enough common elements to make a general correlation possible. The results have been then tested against the various schemes proposed in the last decade and the following notes are the result. The sequence of Australian zones is that given by Bulman (1970, p. V102).

The base of the Ordovician is taken at the base of the Lancefieldian Zone of *Tetragr. approximatus* since there is general agreement that this Zone should be included in the Arenigian, although not yet recognised with certainty in Britain. The succeeding Be1 beds of Australia could probably also be included in the same Zone.

Thomas (1960, p. 9) notes that the 'form considered to be *Didymograptus extensus* in Britain differs from both the American and Australian forms of this species and is probably a variant of *D. nitidus*.' Until the type material is critically re-examined it is therefore unwise to use this zonal species as a basis for correlation over wide areas. The subdivisions of the *extensus* Zone proposed by Elles (1933) have, however, been increasingly used as full zones and it is possible to attempt correlation using them. The lowest subzone of Elles' scheme, that of reclined tetragrapti, has not been generally accepted and it is the succeeding Zone of *Didymogr. deflexus* which can be correlated with the Australian Be2 and Be3. Possibly Be4 can also be included although Thomas (1960) notes the incoming of *Didymogr. nitidus* (Hall) in Be4 which would link it with the British *nitidus* Zone. It is unfortunate that *Tetragr. fruticosus* (Hall) is apparently restricted in Britain to the Girvan area since the division of the Bendigonian relies so much on variation in that species. It is mentioned from the Lake District by Elles & Wood but Jackson (1961, 1962) does not list it.

The incoming of *Didymogr. protobifidus* Elles and other pendent forms was taken by Elles to mark the *nitidus* Zone but Jackson (1962) notes that these do not appear in any numbers until the *Didymogr. hirundo* Zone. He characterises the *nitidus* Zone by the rarity of *Didymogr. deflexus* Elles & Wood and *Didymogr.* aff. *v-fractus* Salter which occur in the zone below and of *Isogr. gibberulus* (Nicholson), characteristic of the zone above. On these criteria, the *nitidus* Zone correlates with the whole of the Chewtonian. It is here that the main discrepancies between the proposed correlations can be seen. The matter is further confused by Thomas (1960) giving only two divisions of the Chewtonian while Harris & Thomas (1938) originally listed three. Their uppermost one (Zone of *Didymogr. balticus*) clearly cannot be correlated directly with the Baltic zone of the same name. The

Great Britain		Australia	
Ashgill	D. anceps		Bolindian
	D. complanatus	D. complanatus	
	P. linearis	P. linearis	
Caradoc	D. clingani	D. hians	Eastonian
	C. wilsoni	C. baragwanathi	
	C. peltifer	C. peltifer / D. multidens	Gisbornian
Llandeilo	N. gracilis	N. gracilis	
	G. teretiusculus	G. teretiusculus M.0.4	
Llanvirn	D. murchisoni	D. decoratus M.0.3	Darriwilian
	D. bifidus	G. intersitus M.0.2	
		G. austrodentatus M.0.1	
Arenig	D. hirundo	Oncograptus / Cardiograptus	Yapeenian
		Oncograptus	
	I. gibberulus	I. caduceus maximodivergens	Castlemainian
		I. c. victoriae	
		I. c. lunata	
	D. nitidus	D. balticus	Chewtonian
		D. protobifidus	
		D. protobifidus / T. fruticosus	
	D. deflexus	T. fruticosus (3-br.)	Bendigonian
		T. fruticosus (3-br. & 4-br.)	
		T. fruticosus (4-br.)	
		T. fruticosus / T. approximatus	
	(T. approximatus)	T. approximatus	Lancefield 3

FIG. 3 A correlation chart of the graptolite zones of Britain and Australia.

occurrence in Norway of *Didymogr. v-fractus* links the Scandinavian zone with the *deflexus* Zone in Britain, that is before *Didymogr. protobifidus* and well before the appearance of *Isogr. gibberulus*. On the other hand, Monsen (1937) records *Isogr. norvegicus* Monsen from the zone immediately above her *balticus* Zone and notes that the form is close to *Isogr. caduceus lunatus* Harris, the index for the base of the Castlemainian. Unfortunately few sequences showing the changes in fauna through continuous sections have been published. Erdtmann (1965) gives one showing *Isogr. gibberulus* appearing before *Isogr. norvegicus* which in turn is followed by *Isogr. lunatus* associated with *Didymogr. hirundo* Salter. If this last record is taken at its face value then the base of the Castlemainian correlates with the base of the *hirundo* Zone which, with *nitidus* Zone equalling the Chewtonian, leaves no space for the *gibberulus* Zone. Elles (1933) and Jackson (1962) characterise the *gibberulus* Zone by an abundance of phyllograptids but these are rare in Australia. It would appear best to correlate the *gibberulus* Zone in Britain with the Castlemainian (with its progression of isograptids) leaving the *hirundo* Zone to start with the Yapeenian, in both of which diplograptids appear in some force.

The absence in most British deposits of *Oncograptus* and *Cardiograptus*, characteristic of the Yapeenian, makes direct correlation difficult but the record of *Oncograptus* from Ireland (Dewey *et al.*, 1970, p. 30) is held to substantiate the equation with the *hirundo* Zone. The position of the *hirundo/bifidus* and the *bifidus/murchisoni* zonal boundaries in the Australian scheme are at present uncertain in the absence in Australia of the tuning-fork didymograptids. The basal Zone of the Darriwilian is that of *Glyptogr. austrodentatus* Harris & Keble. Bulman (1963) has described four varieties of this species which may be essentially geographical races but for which there is some evidence of temporal variation, the British forms occurring in the *gibberulus* Zone and possibly the *hirundo* Zone. Skevington (*in* Dewey *et al.* 1970) attributes one of the Irish faunas to the *bifidus* Zone and also to M.O.2 so that perhaps M.O.1 can be assigned to the *hirundo* Zone. The *bifidus/murchisoni* boundary must remain uncertain until the ranges of the diplograptids in Britain are more precisely known so that they can be matched with the Australian succession.

From the species given by Thomas (1960, chart), a correlation of the Australian Zone of *Glyptogr. teretiusculus* with the British one appears not unreasonable although in the text he stresses the occurrence of *Glossogr. hincksii* (Hopkinson) and *Pterogr. lyricus* Keble & Harris. The latter genus is not known in Britain but is characteristic of the *murchisoni* Zone equivalent in Scandinavia which is there succeeded by the Zone of *Glossogr. hincksii*. This last species was described originally from Britain and is abundant in the *Nemagr. gracilis* Zone so that the Australian *teretiusculus* Zone could span a wide interval. Jaanusson (1960, p. 355) however considers that the topmost Darriwilian (M.O.4) should be correlated with the *murchisoni* Zone in Britain and Scandinavia, leaving M.O.2 and M.O.3 as correlatives of the *bifidus* Zone.

The higher zones have not attracted so much discussion as it has been assumed that Upper Ordovician faunas are less 'provincial'. As Skevington (1970b, p. 561) has pointed out, however, the differences remain but at a specific rather than generic level. It is also possible that the Upper Ordovician graptolite sequence, which in Britain is primarily based on continuous sections, has not been so minutely subdivided since it is based on Lapworth's original work and the broader zones may be more easily compared internationally.

The incoming of *Dicellograptus* and *Nemagraptus* (taken as indicating the Gisbornian) should provide a key horizon but recent work in South Wales (Toghill 1970a) has demonstrated *Dicellograptus* from the *teretiusculus* Zone as well as a late *Isograptus*. Slender *Nemagraptus* from the same Zone at Builth confirms the view that there is a gradual incoming of both genera and that the easily recognisable *Nemagr. gracilis* (Hall) may not be exactly contemporaneous throughout its world-wide distribution. With this qualification, the base of the Gisbornian can be matched with the *gracilis* Zone and its upper part with the mixed fauna usually assigned to the *Climacogr. peltifer* Zone. The succeeding Eastonian includes both *Climacogr. wilsoni* and *Dicranogr. clingani* Zones. Elles claimed that *Cimacogr. baragwanathi* Hall was a synonym of *Climacogr. wilsoni* Lapworth but this has not been critically established. Thomas (1960) includes *Leptograptus* spp. in his Eastonian lists and Öpik (1958) suggests that *Pleurograptus* may be present at Canberra in Eastonian beds so it is possible that the base of the Bolindian may sometimes be taken higher than the *clingani/linearis* boundary in Britain. Generally speaking the lower Bolindian fauna appears to be that of the *Pleurogr. linearis* Zone while the upper one combines elements of both *Dicellogr. complanatus* and *Dicellogr. anceps* Zones of Britain. This is a usual feature of the uppermost Ordovician graptolite faunas. In all cases there is a profusion of diplograptids leading into a transition to the basal Silurian faunas and it is the lingering dicellograptids which provide the essentially Ordovician element. In their absence, determination of the system boundary is very difficult, if not impossible.

In the lower part of the Bolindian Thomas lists several diplograptids such as *Climacogr. bicornis* (Hall) and *Climacogr. caudatus* Lapworth, which in Britain are limited to lower horizons, and it is clear that much further work on the identification and ranges of species in the Upper Ordovician is required.

3. Wales

By DOUGLAS ANTHONY BASSETT

THE disproportionate attention paid to the Ordovician rocks of Wales, when compared to other areas in the British Isles, is, in large part, due to Charles Lapworth's recognition of the System as 'Strata included between the base of the Lower Llandovery formation and that of the Lower Arenig' (1879a, p. 14) on the basis of sections in North Wales and with particular reference to 'the whole of the great Bala district'. Furthermore this is where Adam Sedgwick 'first worked out the physical succession among the rocks of the intermediate or so-called *Upper Cambrian* or *Lower Silurian* system'. The recognition of this succession as a standard stimulated research into regional relationships and resulted in the publication of a very large number of papers, monographs, memoirs and maps (Bassett 1961, 1963, 1967).

As a result of these investigations the type areas of the three lower Series and of five of the six constituent graptolite zones are in the Principality, and the uppermost stage of the Ashgill is based on the shelly succession in the Hirnant valley south-east of Bala. The 'Bala Series' recognized by Lapworth as the type for the Upper Ordovician has gradually been superseded by the Caradoc and Ashgill Series, and, as noted in the Introduction, it has recently been suggested that because of the discovery of a major break in the sequence at Bala, the use of the term 'Bala', as a Series name, be discontinued.

There is no single modern review of the stratigraphy of the Welsh Ordovician, although the rocks of the Principality figure prominently in most of the reviews of the Ordovician of the British Isles (*see*, in particular, O. T. Jones 1936, 1938, 1956; A. Williams 1969b). The geological map of Wales compiled by Jones (1938, Pl. A) is still the only published map which depicts the generalised areal distribution of the various Series of the System.

Orthography of Welsh place-names. Although the spelling of very many of the Welsh place-names in general use in stratigraphical literature has been modified recently, the more familiar renderings are retained in this paper. The basis of the revision, which is clearly reflected in the current maps of the Ordnance Survey, is discussed in *A gazetteer of Welsh place-names* (University of Wales Press, Cardiff. 1957).

Correlation of the Ordovician of Wales, excluding Welshpool and the Breidden Hills, is indicated in columns W 1 to W 12 (Fig. 6); the distribution of the outcrops and the extent of the areas surveyed in detail in Wales and the Borderland in Figs. 5 and 4 respectively.

(A) ANGLESEY (Column W 1)

The Ordovician is represented by a very varied sequence of Arenig, Llanvirn, Llandeilo (?) and early Caradoc age, attaining a maximum thickness of approximately 1,600 m, over half of which is accounted for by the Arenig.

The graptolitic facies was described in detail by Greenly (1919) and Bates, in a recent resurvey, has concentrated on the shelly facies. The stratigraphy is now known to be more complicated than that envisaged by Greenly and, because of the difficulties of precise lithological correlation, Bates has erected over thirty new lithostratigraphic units.

In the largest of the seven separate outcrops ('Principal area' [W 1]), where the succession is more complete than elsewhere, the grits and sandstones resting on the Mona Complex contain a shelly fauna which, although meagre in trilobites, contains the most varied brachiopod assemblage in the Arenig of Wales (Bates 1968; 1969a, pp. 155, 157). Some of the brachiopods, particularly *Lenorthis proava* (Salter), occur in very large numbers. The closest correlation is with the Tagoat Beds of Co. Wexford (Baker 1966, pp. 2–4) which are of Lower Arenig age (Bates *in* Brenchley, Harper & Skevington 1967). As with younger shelly faunas, the affinities are Baltic. Graptolite faunas of the *Didymogr. extensus* and *Didymogr. hirundo* Zones have also been recorded (Bates 1972, pp. 36–8).

Llanvirn shelly faunas are sparse and the sediments are predominantly shaly. Graptolite faunas of *bifidus* and *murchisoni* age have been recorded (Bates 1968, Fig. 2; Greenly 1919, p. 459, etc.; Hawkins 1966, p. 13).

Fossils attributed to the *Glyptogr. teretiusculus* Zone have been found associated with an ironstone at Bonw (Greenly 1919, p. 465) and in a borehole at Llangoed

(Greenly 1919, pp. 432, 433) but the extent and thickness of the Zone is not known.

The ironstone and shales at Llanbabo (Fferam Ironstone and Fferam Shales of Bates 1968, 1972) have yielded a rich graptolite fauna of the *Nemagr. gracilis* Zone (Elles *in* Greenly 1919, p. 453). In the overlying Llanbabo Church Grits (Bates 1968, p. 136) graptolites from the interbedded shales indicate a position high in the same Zone (Elles *in* Greenly 1919, p. 455). The grits of the formation have yielded a shelly fauna (Bates 1968, p. 137; 1972, p. 53) which correlates well with the faunas in the Tandinas Shales at Caregonen and the Derfel Limestone at Arenig (Whittington & Williams 1955) although no trilobites are recorded at Llanbabo. Stocks occurring at Llanbabo, but not so far found at Derfel, provide further evidence of the Baltic affinities of the fauna.

The prolific fauna in the limestone blocks in the spectacular breccia (Garn Formation) developed in the *Nemagr. gracilis* Zone at Porth Padrig (Greenly 1919, vol. 2, p. 478; Bates 1972, p. 51) is not younger than *gracilis* and may be older (Bates 1968, p. 142). The trilobites of the limestone are similar to those of the Derfel Limestone but the brachiopods are somewhat 'ambiguous'.

Graptolites from the *Climacogr. wilsoni* and *Dicranogr. clingani* Zones have been recorded at Llanbabo (Elles *in* Greenly 1919, pp. 454–5), but there is no evidence of the extent of the Zones.

FIG. 4 Index map of the areas of Ordovician rocks in Wales and the Welsh Borderland which have been surveyed since 1890.

1. Greenly (1919); 2. Bates (1964); 3. Matley (1928); 4. Matley (1932); 5. Nicholas (1916); 6. Fitch (1967); 7. Matley (1938); 8. Matley & Heard (1930); 9. Tremlett (1962); 10. Tremlett (1965); 11. Tremlett (1964); 12. Harper (1956); 13. Roberts (1967); 14. Fearnsides (1910); 15. Fearnsides & Davies (1944); 16. Jennings & Williams (1891); 17. Bromley (1965); 18. Beavon (1963); 19. Rast (1961); 20. Shackleton (1959); 21. Williams, H. (1927); 22. Cattermole & Jones (1970); 23. Hughes (1917); 24. Elles (1904); 25. Greenly (1944); 26. Davies, R. A. (1969); 27. Evans, C. D. R. (1968); 28. Williams, D. (1930); 29. Ramsay (1959); 30. Williams & Bulman (1931); 31. Williams, H. (1922); 32. Diggens & Romano (1948); 33. Ravies, D. A. B. (1936); 34. Stevenson (1971); 35. Elles (1909); 36. Lynas (1970); 37. Fearnsides (1905); 38. Wells (1925); 39. Cox & Wells (1921); 40. Cox (1925); 41. Davies, R. G. (1959); 42. Jones, B. (1933); 43. Jehu (1926); 44. Jones & Pugh (1916); 45. Pugh (1923); 46. Pugh (1928); 47. Pugh (1929); 48. Bancroft (1928); 49. Elles (1922); 50. Bassett *et al.* (1966); 51. Blackie (1928); 52. Smith (1935); 53. Wills & Smith (1922); 54. Groom & Lake (1908); 55. Wedd *et al.* (1927); 56. Brenchley (1966); 57. Jones, G. H. (1956); 58. MacGregor (1958); 59. King (1923); 60. Whittington (1938); 61. Wedd *et al.* (1929); 62. King (1928); 63. Cave (1965); 64. Cave (1955); 65. Wade (1911); 66. Whittard (1932); 67. Hains (1969); 68. Wright (1968); 69. Hains (1970); 70. Cobbold (1927); 71. Stubblefield & Bulman (1927); 72. Pocock & Wray (1925); 73. Pocock *et al.* (1938); 74. Jones, O. T. (1922); 75. Jones, O. T. (1909); 76. James (1971); 77. Jones, W. D. V. (1945); 78. Roberts (1929); 79. Lapworth (1900); 80. Davies, K. A. (1928); 81. Davies, K. A. (1926); 82. Davies, K. A. (1933); 83. Drew & Slater (1910); 84. Evans, D. C. (unpublished map, *see* Jones, O. T. 1938); 85. Anketell (1963); 86. Hendricks (1926); 87. Williams, A. (1953); 88. Potter (1960); 89. Jones, O. T. (1925); 90. Jones, O. T. (1949); 91. Stamp & Wooldridge (1923); 92. Andrew (1925); 93. Jones, O. T. (1947); 94. Jones & Pugh (1949); 95. Elles (1940); 96. Crosfield & Skeat (1896); 97. Cantrill & Thomas (1906); 98. Evans, D. C. (1906); 99. Evans, W. D. (1945); 100. Reed (1895); 101. Thomas & Thomas (1956); 102. Cox (1916); 103. Waltham (1971); 104. Williams, T. G. (1934); 105. Thomas & Cox (1924); 106. Thomas & Jones (1912); 107. Green (1908); 108. Pringle (1930); 109. Cantrill *et al.* (1916); 110. Strahan *et al.* (1914); 111. Strahan *et al.* (1909); 112. Strahan *et al.* (1907); 113. Dixon (1921).

The titles of the publications are given in Section 9 (References).

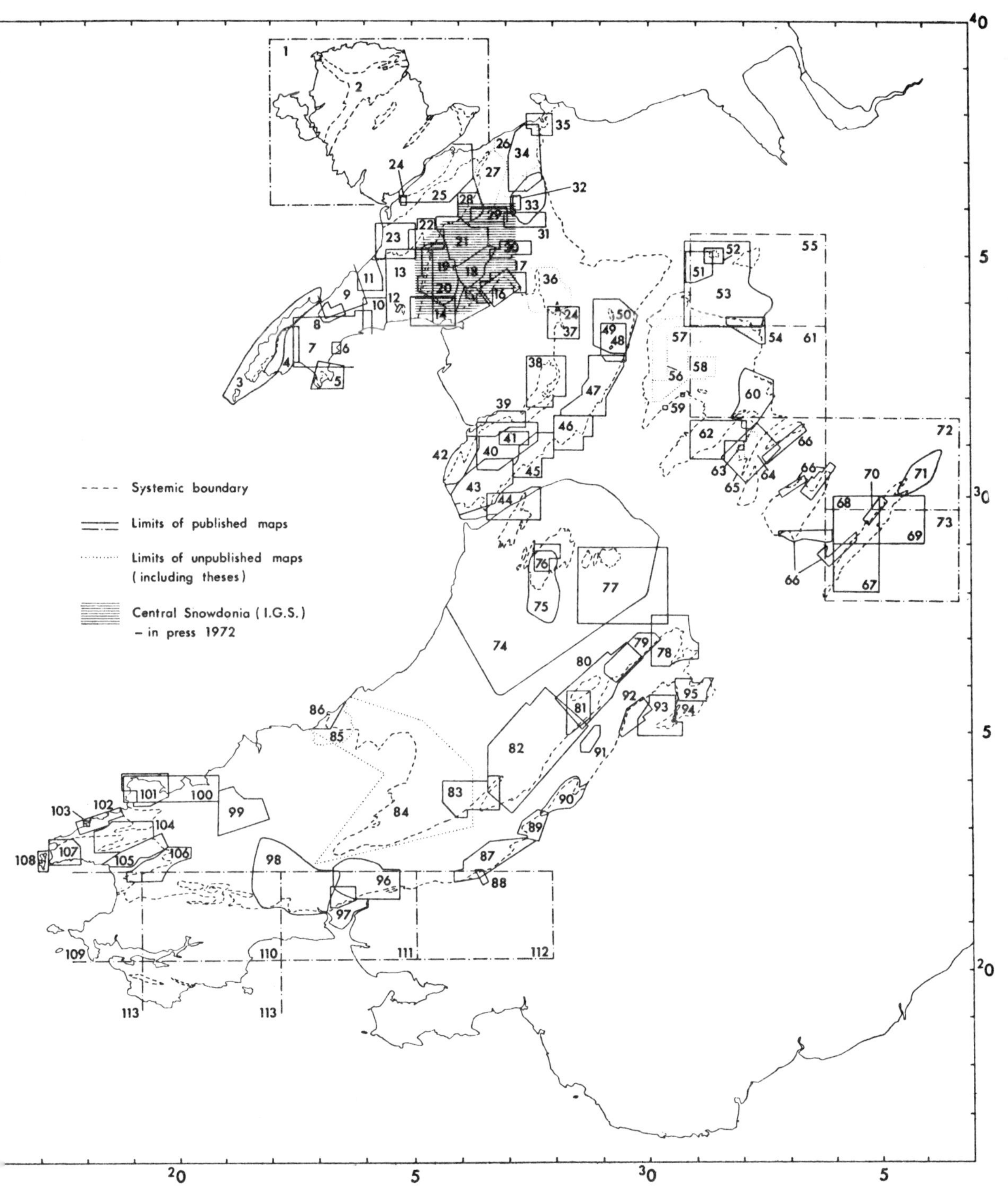

Systemic boundary
Limits of published maps
Limits of unpublished maps
(including theses)
Central Snowdonia (I.G.S.)
— in press 1972

(B) ARVON
(not included in the chart)

Ordovician rocks on the mainland north of the Padarn ridge in Caernarvonshire occur in one relatively large tract, extending without interruption from Penrhyn Castle to beyond Caernarvon, and seven small ones (Greenly 1944, Fig. 1). Each tract was interpreted by Greenly (1944, Fig. 4) as an asymmetrical infold with its south-western limb ruptured, but D. S. Wood (1969) has since suggested a much simpler structural style.

The overall facies of the rocks is similar to that on Anglesey. Rather more than 750 m of strata (Greenly 1944, p. 79), made up predominantly of shale, with a persistent unnamed grit at the base and a distinctive lenticular ironstone near the top (Pulfrey 1933), contain fossils referable to the *extensus, hirundo, bifidus* and *teretiusculus* Zones (Greenly 1944, pp. 77, 78). The apparent absence of a *murchisoni* fauna is considered to be the result of poor exposures. The estimated thickness may well be considerably in error because the vertical dimensions of the strata in the Bethesda slates have been nearly doubled (Sorby 1908, p. 221; Dr. D. S. Wood, personal communication).

The best known section is in Afon Seiont at Caernarvon where the three lowest zones of the system are clearly exposed (Elles 1904, Fig. 1) and from which both trilobites and graptolites have been recorded. Strata of the *extensus* Zone appear to be faulted against the Pre-Cambrian.

It was on the basis of this and other North Wales sections that Elles (1904), preferring to stress the abundance of extensiform species of *Didymograptus* in beds below those yielding a *bifidus* Zone fauna, substituted the two zones of *Didymogr. extensus* and *Didymogr. hirundo* for Lapworth's *Tetragraptus* Zone erected earlier (1880, p. 197) on the basis of sections in South Wales.

(C) LLEYN, SNOWDONIA AND CONWAY (Columns W2 to W4)

Since the Second World War the well-known studies in Snowdonia (H. Williams 1922, 1927; D. Williams 1930; Williams & Bulman 1931; D. A. B. Davies 1936) and Lleyn (Matley 1928, 1932, 1938; Matley & Heard 1930; Matley & Smith 1936) have been supplemented and modified by a great deal of detailed mapping. This has included work by a number of individuals (Beavon 1963; Bromley 1965; Cattermole & Jones 1970; Davies 1969; Diggens & Romano 1968; Evans 1968; Fearnsides & Davies 1944; Fitch 1967; Lynas 1970; Rast 1961; Roberts 1967; Shackleton 1959; Stevenson 1971; Tremlett 1962, 1964, 1965) and by officers of the Institute of Geological Sciences (Francis *et al.* 1970). The emphasis has undoubtedly been on the igneous and structural geology and the most striking single advance has almost certainly been the recognition of ignimbrites (Oliver 1954) and the extent of their distribution (Rast *et al.* 1958).

There is, as yet, no single written assessment of these advances but a number of reviews of restricted aspects have been published (Bassett 1969; Brenchley 1969; Rast 1969; Tremlett 1969). The 2½ inch-to-the-mile (1 : 25,000) geological map now in press (Institute of Geological Sciences 1972) does, however, provide a valuable synthesis.

Detailed mapping has resulted in the recognition of a very large number of new lithostratigraphic units and the modification of many others, but the discovery and recognition of horizons of biostratigraphic significance has not been anything like as spectacular. Enough is, however, now known of the faunas to warrant a consideration of each Series as a separate entity.

The thickness of the System in Snowdonia has been estimated as between 2,400–3,000 m on the basis of measurements at section. Allowing for tectonic dilation, however, the original thickness was probably between 1,800–2,100 m (Rast 1969, p. 329) and is thus comparable to that in the Shelve district.

The base of the Ordovician. Throughout the greater part of Caernarvonshire, Ordovician strata rest with pronounced unconformity on the Cambrian and Pre-Cambrian with the magnitude of the break decreasing steadily south-eastward as far as Tremadoc (Shackleton 1954, p. 268, Fig. 1; Jones 1938, p. lxxii; Cowie *et al.* 1972, Pl. 2) where the youngest known pre-Arenig beds of the district are preserved. The exact nature and position of the junction is not known for certain along the northern limb of the Snowdon syncline: the contact appears to be slipped or faulted at Moel Hebog (Shackleton 1959, p. 224); the unfossiliferous Plas-y-Nant Beds (Slates) of Snowdon have been attributed to the Ffestiniog Group (Shackleton 1959, p. 222) and the Arenig Series (H. Williams 1927, p. 354); and the contact may be conformable east of Bethesda (Evans 1968, p. 22).

At Tremadoc the unfossiliferous basal grit of the Arenig Series rests without any apparent unconformity (although there is some evidence of an erosional contact, Fearnsides 1910, p. 164) on the Garth Hill Beds. These are considered (Cowie *et al.* 1972, p. 11) to represent the uppermost, or *Angelina sedgwickii* Zone of the Tremadoc Series.

Arenig and Llanvirn Series. The base of the Arenig Series is marked throughout the area by the thin impersistent Garth Grit and its equivalents (Jennings & Williams 1891, p. 373; Hughes *in* Fearnsides 1905, p. 639; Fearnsides & Davies 1944, p. 252). Undoubted fossil remains are absent but the problematical *Bolopora undosa* Lewis is generally considered to be a good guide to this horizon (Lewis 1926). It has been recorded in the basal grit at a number of localities in Lleyn and Snowdonia (e.g. Lewis *in* Fearnsides & Davies 1944, p. 274; Evans 1968, p. 24; Davies 1969, pp. 14, 15). Recently, however, examples have been collected from the uppermost 6 m of the Tremadoc Series underlying the Garth Grit near Ffestiniog and also in grits up to 30 m above (Bates 1969a, p. 157). Recent mapping of the type locality of the formation (Lynas 1970) suggests that the contact between the Grit and the underlying Tremadoc shales is gradational and not affected by a series of thrusts as originally maintained by Fearnsides (*in* Lewis 1926). Almost immediately to the east of the type-locality, but across a major northerly trending fault, the relationship is an unconformable one. Lynas suggests that contemporaneous movements along the fault influenced sedimentation.

The two zones of the Arenig are clearly differentiated in south-western Lleyn (Matley 1928, p. 493; 1932, pp. 259–61) but not in the remainder of the peninsula (Fitch 1967; Matley 1938; Nicholas 1915; B. Roberts 1967). The subdivisions of the *extensus* Zone erected by Jackson in the Skiddaw Slates (1962) have not yet been recognised in North Wales successions. Some kind of sequence is, however, apparent in Lleyn in that the commonly occurring *Azygogr. eivionicus* Elles is considered (Elles 1922b, p. 299) to be characteristic of the middle of the Zone and referred to as a subzonal fossil (Matley 1932, p. 259). This is in general agreement with the findings in Cumberland where the azygograptids appear somewhere near the top of the *deflexus* Subzone, where *Azygogr. eivionicus* occurs locally in great numbers, and where the genus has its peak in the *Didymogr. nitidus* Subzone where it is represented by the three species *eivionicus*, *lapworthi* and *suecicus*. In the Mynydd Rhiw and Sarn district, at least, the occurrence of *Azygogr. suecicus* Moberg is taken (Matley 1932, pp. 248, 253) as indicating the *hirundo* Zone. In contrast to the sequence in Cumberland, phyllograptids are rare in north-west Wales and those that occur do so in a thin band in the *hirundo* Zone of south-west Lleyn (Matley 1928, p. 488) and Anglesey.

A number of facies, defined with particular reference to lithology and to biogenic and inorganic sedimentary structures, are recognised in the Arenig of south-western Lleyn (Crimes 1970) and an isopachyte map of the *extensus* Zone sediments in North Wales generally, based in part on unpublished work, is available (Crimes 1970, Fig. 5).

The two Zones of the Llanvirn are not clearly differentiated anywhere in the area although graptolite faunas from both Zones occur frequently. At Moel Hebog four apparently distinct faunas occur (Shackleton 1959, pp. 226, 227) but their stratigraphical relations are not known. They are characterized respectively by *Didymogr. euodus* Lapworth, *Didymogr. geminus* (Hisinger), an early divergent and a late form of *Didymogr. murchisoni* (Beck). A cyclopygid shelly fauna has also been recorded (Roberts 1967, p. 374).

Llandeilo (?) Series. It is still not clear whether the *Glyptogr. teretiusculus* Zone of the Llandeilo Series is represented. The Zone fossil is recorded in many parts of Caernarvonshire (*see* list of references in Shackleton 1959, p. 228; *see also* Sargent 1924, p. 84), but its occurrence alone is not necessarily diagnostic of the Zone.

Caradoc Series. In all the published literature on Snowdonia the base of the Caradoc Series is taken at, or a little below, the incoming of *Nemagr. gracilis*. As a result of Addison's work, however, (*see* p. 36) it is conceivable that some of these occurrences (e.g. H. Williams 1927, p. 359; Nicholas 1915, p. 123; Shackleton 1959, p. 231, Fearnsides 1910, p. 170) are Llandeilo in age. The suggestion that the fauna found in the tuffs near the base of Y Glog Volcanic Group is from near the top of the *gracilis* Zone or even a little higher is confirmed by the discovery of a shelly fauna closely comparable to that of the Derfel Limestone in the upper tuff horizon of the laterally equivalent Moelwyn Volcanic Group (Bates *in* Bromley 1965, p. 254).

These early Caradoc (or Llandeilo ?) volcanics are separated from the main Snowdon Volcanic Group by the widely distributed Glanrafon Beds. The evidence provided by the shelly faunas for the age of these sediments, and for the commencement and cessation of the subsequent volcanism, has recently been reviewed (Diggens & Romano 1968, Pl. 3; Brenchley 1969, pp. 185–7, Fig. 36;—with minor emendations in Stevenson 1971). Briefly, the combined evidence is of probable Harnagian faunas from two localities only (Harper *in* Shackleton 1959, p. 249; Dean *in* Stevenson 1971, p. 530) and undoubted Soudleyan and Longvillian faunas from a number of localities (Dean 1965; Dean *in* Stevenson 1971; Diggens &

Romano 1968; Diggens & Romano *in* R. A. Davies 1969, pp. 65, 66; Harper 1947, 1956; Harper *in* Shackleton 1959; Harper *in* Tremlett 1962; Lamont 1946; Stubblefield *in* Matley 1938; Williams & Harper *in* Beavon 1963; Wood & Harper 1962).

Differentiating the faunas of the upper and lower divisions of the Longvillian of North Wales on the basis of the faunas described from south Shropshire has proved to be difficult. It may therefore be advisable to follow Whittington (1962-68, pp. 93-138) and Romano & Diggens (1969) in correlating North Wales sections with the sequence at Bala and to classify the Longvillian faunas as pre- and post-Cymerig ('Bala') Limestone (*see* p. 25). Difficulties of another kind arise in correlating the brachiopods. Because the speciation of the dalmanellids and sowerbyellids is now based largely on statistical methods (Williams 1963) it has been found difficult to obtain a satisfactory statistical comparison with, for example, the faunas in Dolwyddelan (Romano & Diggens 1969, p. 603) because of the deformed nature of the specimens and the commonly imperfect preservation.

The sediments overlying the volcanic sequence are mainly graptolitic. At Conway (Strachan *in* Wood & Harper 1962) and Dolwyddelan (Williams & Bulman 1931, p. 444) the black slates have been referred to the *Dicranogr. clingani* Zone 'or only very slightly lower' and 'the higher part of the zone of *D. clingani*', respectively. In the river Dwyfach near Llanystwmdwy, black flaggy mudstones contain orthograptids belonging to the *clingani* Zone and are overlain by black shales containing an abundant fauna belonging to the *Pleurogr. linearis* Zone (Harper 1956, pp. 389, 390). In the area around Pwllheli, on the other hand, Elles (*in* Matley 1938, p. 593) records three different assemblages from beds immediately overlying the volcanics—from the *multidens*, basal *clingani*, and uppermost *clingani* (or possibly basal *linearis*) Zones respectively.

Cave's (1965) suggestion that many of these black shales should be referred to the Nod Glas is questioned by Whittington (1962-68, p. 112) who points out that near Llanystwmdwy (Harper 1956, p. 389) these shales are underlain by rocks only yielding *Estoniops alifrons* (M'Coy) and which therefore indicate a horizon above the Cymerig ('Bala') Limestone.

The sequence at Penarwel originally described by Matley (1938) has since been reinterpreted (Fitch 1967, Fig. 1, Table 1). The volcanic strata (renamed the Llanbedrog Ignimbrite Group) are succeeded conformably by the Penarwel Drive Beds which contain a shelly fauna of Upper Longvillian age and a graptolite fauna attributed to the *Diplogr. multidens* Zone. These beds are overlain unconformably by a black slate formation with a local conglomerate at the base which contains a graptolite fauna indicative of the top of the *clingani* Zone and correlated with the Nod Glas horizon.

In the essentially graptolitic Caradoc sequence at Conway Elles (1909) recognised faunas which probably represented the *peltifer* and *wilsoni* Zones as well as an intermediate assemblage which she referred to a Zone of '*Mesograptus multidens* and *Dicranograptus brevicaulis*'. On the basis of a re-examination of much of the fauna Strachan suggests (*in* Wood & Harper 1962, p. 185) that there is no apparent justification for such an elaborate zonal sub-division at Conway.

Ashgill Series. Ashgillian strata occupy a very limited area in the region. The

Crugan Mudstones which overlie the black shales at Penarwel have a fauna which is very like that of the Rhiwlas Limestone (Matley 1938, pp. 592, 593; Whittington 1962-68, Table 4). Unnamed beds in the Dwyfach and Dwyfor sections near Llanystwmdwy (Harper 1956, pp. 390–3; Roberts 1967, p. 378) yield a fauna similar to the Crugan Mudstones. They also contain a graptolite fauna very tentatively assigned (Bulman *in* Harper 1956, p. 391) to the *Dicellogr. complanatus* Zone. These are overlain in turn by unfossiliferous Ashgillian strata in faulted contact with Llandovery shales containing a fauna attributable to the *Monogr. crispus* and *Monogr. fimbriatus* Zones.

The fullest lithological sequence of the Ashgill Series is in north-eastern Caernarvonshire where the Bodeidda Mudstones and the Deganwy Mudstones of Conway (Elles 1909) overlie the Cadnant Shales with apparent conformity. There is a distinct break in the sequence, however, because the overlying Bodeidda Mudstones contain a shelly fauna which probably equates with the upper Cautleyan or lower Rawtheyan of the Cautley sequence. If the presence of *Anisopleurella quinquecostata* (M'Coy) is significant, it may indicate Zone 5 of Ingham's sequence (1966, p. 501).

The overlying Deganwy Mudstones contain *Orthogr. truncatus abbreviatus* Elles & Wood in the lower part of the sequence—a species considered by Toghill (1970b, p. 24) to be restricted to the *Dicellogr. anceps* Zone, which in turn equates approximately with the Rawtheyan (Ingham & Wright 1970, Fig. 1, p. 236). The presence of *Dalmanitina mucronata* (Brongniart) in the upper part of the formation, however, indicates a Hirnantian age. The overlying Conway Castle Grits contain *Eostropheodonta hirnantensis* (M'Coy) which also indicates a Hirnantian age. Apparent lateral equivalents of the grit, however, contain a characteristic Lower Llandovery graptolite fauna: Cocks *et al.* (1971, Fig. 2, col. 4) place the Conway Castle Grits in the Llandovery with the proviso (p. 110) that there are 'conflicting faunal reports; they [the grits] may be partly or even wholly, of Ashgill age'.

(D) EASTERN FLANKS OF THE HARLECH DOME (Columns W5 and W6)

The greater part of the Ordovician sequence can be traced in an almost unbroken outcrop for over 40 miles from Snowdonia, around the eastern flanks of the Harlech Dome, to the coast between Ffairbourne and Aberdovey. In contrast to Lleyn and Snowdonia the strata here are not greatly affected by folding, and only locally are they disrupted to any great extent by faulting. The sequence differs from the Caernarvonshire one in that: (i) there is a considerable development of Arenig-Llanvirn volcanics; (ii) that the Soudleyan-Longvillian vulcanicity so spectacularly developed in Snowdonia is represented by occasional thin tuff bands; (iii) that the Ashgillian is well exposed throughout the belt. As in Snowdonia, however, the position regarding the Llandeilo is problematical. Another contrast is that there has been proportionally much less work carried out since the Second World War in Merioneth and much the greater part of our knowledge is therefore based on a small number of well-known studies (Cox 1925; Cox & Wells 1921; Fearnsides 1905; Jehu 1926; Pugh 1923, 1928, 1929; Wells 1925).

Arenig–Bala (W5)

From a biostratigraphical standpoint the most appropriate section to take is that from the western flanks of Arenig Fawr mountain through the town of Bala to the upper reaches of the Hirnant valley south-east of Bala Lake (Llyn Tegid). Historically, it was here that Sedgwick (1838, 1845, 1852) defined the Arenig and Bala Groups. It was here also that he recognised and mapped the three distinctive limestones—Bala, Rhiwlas and Hirnant—which have played such an important role in the correlation of the upper part of the System. Sedgwick used them as the basis for the subdivision of his Bala Group and it is likely that Marr's choice (1905) of the base of the Ashgill was influenced by his understanding of the sub-divisions of the Bala section. Two of the names are still valid, but the term 'Bala Limestone', being a pre-occupied group name, has been replaced by the term 'Cymerig Limestone' (Bassett *et al.* 1966, p. 227). As mentioned earlier (*see also* Whittington & Williams 1964; and A. Williams 1969b), the area was undoubtedly the one chosen by Lapworth (1879a) as the type section of his new System. And it was here that Bancroft (1928) first put forward correlative subdivisions of his Shropshire succession and in so doing revealed the presence of a pre-Rhiwlas unconformity; and where he also erected the Hirnantian Stage for rocks which contain the youngest Ordovician shelly fauna in Britain.

The entire Ordovician sequence consists of about 3,660 m of comparatively well-exposed sedimentary, pyroclastic and volcanic rocks which succeed Tremadoc shales unconformably and which are, in turn, overlain conformably by graptolitic shales of Llandovery age. More than half the succession is nearly or completely unfossiliferous. Nevertheless, with the exception of the middle parts of the sequence, enough key graptolitic and shelly faunas have been found to enable reasonably accurate correlation to be made with successions elsewhere in Britain. The recent studies of the Caradocian trilobites (Whittington 1962–68) and the quantitative studies of the brachiopods (A. Williams 1963) demonstrate, however, that the Zones of the Shropshire Caradoc (Dean 1960a, 1961, 1963a, etc.) cannot be recognised and that even the use of Bancroft's Stages calls for a number of compromises. The problem arises because only 23 of the 52 brachiopod genera recovered from the Caradocian of Bala and Shropshire are found in both regions, and only 5 of these are known to be represented by conspecific forms. Certain trilobites appear to be more widely distributed but, nevertheless, conspecific forms number less than one-third of the 31 species listed from the Lower Bala Group. The differences in the brachiopod faunas of the two areas appear to be accentuated by the existence of three distinct assemblages within the successions at Bala, which suggest environmental control.

The details in the column (W 5) are based on Fearnsides (1905), with a minor modification by Whittington (1966), and Bassett *et al.* (1966).

Arenig Series. In terms of the stratigraphical succession given in the first detailed account of the Arenig district (Fearnsides 1905, p. 609), Sedgwick conceived the Series as extending from the Basal Grit, resting unconformably on the Amnodd Shales of the Tremadoc Series (*Shumardia pusilla* Zone—Cowie *et al.* 1972) to the base of the Derfel Limestone. Fearnsides, however, adopted the definition of the Arenig given by Hicks (1875) and Lapworth (1880), i.e. to include the *extensus, hirundo* and *bifidus* Zones—each of which is recognised in the sequence.

In the current definition of the term, the Arenig sediments at Arenig consist of about 213 m of sedimentary and pyroclastic rocks which yield both shelly and graptolitic assemblages. The faunal list given by Fearnsides (1905, p. 619) together with material in Skevington's collection (Skevington 1969, p. 168) show that the Llyfnant Flags at the base of the Series belong to the lowest part of the *extensus* Zone—the *deflexus* Subzone (Jackson 1962, p. 305). There is, however, no evidence for the underlying *Tetragr. approximatus* Zone.

The overlying Henllan (or *Calymene*) Ash contains (in the constituent Erwent (or *Ogygia*) Limestone) the only shelly fauna from the type area. It correlates closely with the lower third of the Mytton Flags of west Shropshire (Whittington 1966, p. 493) which are regarded (Whittard 1955-67, pp. 16–17, 244, 236) as of *extensus* Zone age.

The upper limit of the Series is clearly marked because the *hirundo* fauna (of the Filltirgerig Beds) dies out suddenly within 30 cm of beds containing *Didymogr. bifidus* (Fearnsides 1905, p. 622).

Llanvirn Series. The Llanvirn is represented by the *bifidus*-bearing Olchfa Shales, the unfossiliferous Lower Ashes and Agglomerate, and the very variable Daerfawr Shales which yield diplograptids and dicellograptids indicative of a position high in the *murchisoni* Zone, although in the absence of the index species, there is no conclusive evidence of this.

Llandeilo (?) Series. It has been claimed (Williams 1969b, p. 244) that both Llanvirn-Llandeilo and Llandeilo-Caradoc boundaries can be identified tentatively in this sequence. The former may be within the Daerfawr Shales, the latter at the base of the Nant Hir Mudstones (Shales). No fauna of undoubted Llandeilo age is, however, known from the area. It is also claimed (Rast 1969, p. 318) that recent mapping of Arenig Fawr by J. N. Diggens suggests strongly that rocks of Llandeilo age are missing.

Caradoc Series. The lowest fossiliferous horizon attributed to the Caradoc is the sporadic Derfel Limestone at the base of the Nant Hir Mudstones and which rests directly on the Upper Rhyolitic Ashes. Much of this rich fauna (Whittington & Williams 1955) is as yet unknown in the Welsh borderlands and correlation is therefore somewhat controversial (Dean 1960b, Spjeldnaes 1957). On the evidence of the trilobites, and particularly the presence of *Broeggerolithus* cf. *harnagensis* (Bancroft), the limestone would appear to be of early Harnagian age or to equate with the boundary between rocks yielding *Nemagr. gracilis* and overlying Harnagian strata of the *Diplogr. multidens* Zone. The occurrence of the distinctive *Kullervo-Palaeostrophomena-Nicolella* fauna in association with *Nemagr. gracilis* and other diagnostic graptolites in Anglesey (Bates 1968) is, however, taken to indicate a Costonian age (Williams 1963, p. 339). The graptolite fauna in Anglesey is similar to that found in the basal mudstones of the Nant Hir Mudstones (Elles 1922a, pp. 144–5) and in their contiguous stratigraphical equivalents, the basal Ceiswyn Slates (Pugh 1929, p. 250)—at or near the horizon of the Derfel Limestone.

The next fossiliferous horizon, near the middle of the Nant Hir Mudstones, is, on the basis of the occurrence of *Broeggerolithus ulrichi* (Bancroft), probably late Harnagian in age (Bassett *et al.* 1966, pp. 229, 257; Dean 1958, p. 202; 1960a,

FIG. 5—The d

higher part of the Ordovician succession. In the absence of faunal horizons, however, the extent of this break is not known.

The only other fossiliferous horizons in the Upper Bala Group are in the Foel y Ddinas Formation, and particularly in the constituent Hirnant Limestone. The formation has yielded *Dalmanitina mucronata* (Brongniart), usually regarded as indicating a late Ashgillian age, though its known stratigraphical range in Europe is much greater (Temple 1952; Kielan 1960, p. 8). The brachiopods *Eostropheodonta hirnantensis* (M'Coy), *Hirnantia sagittifera* (Davidson) and ? *Plectatrypa* sp. are present in the lower beds, and are also the most striking elements of the unusually distinctive fossil assemblages of the overlying Hirnant Limestone and associated mudstones and sandstones. *H. sagittifera* is also known from the basal Llandovery of Haverfordwest (Williams 1951, p. 129) but it has been suggested (Bassett *et al.* 1966, p. 262) that the Foel y Ddinas fauna generally can be taken as the youngest Ordovician shelly assemblage known in Britain. The section at Foel y Ddinas is now taken as the type for the Hirnantian Stage (Ingham & Wright 1970, p. 238 *et seq.*).

The Foel y Ddinas Mudstones are succeeded directly by the Cwm yr Aethnen Shales which contain graptolites from the basal zone of the Llandovery Series (Pugh 1929, p. 274). The Silurian-Ordovician boundary is therefore defined unequivocally in terms of shelly and graptolitic facies.

Cader Idris-Corris (W 6)

Knowledge of the area between Bala and Towyn adds comparatively little to the general biostratigraphical picture. Consequently, in spite of the considerable lithological variation, particularly in the Lower Ordovician, one section only is illustrated in the Chart. It .is appropriate, however, to mention a number of points which need clarification.

(i) The relationship of the Rhobell Fawr Volcanic Group to the base of the Ordovician is not clear. Wells (1925, p. 491) maintained that the formation is post-Tremadoc in the sense that 'it is later than any part of that series which survived the post-Tremadoc erosion in the immediate neighbourhood, and possibly anywhere around the dome, and pre-Arenigian in the sense that it is succeeded unconformably by what is, in all other parts of southern Merioneth, the local base of the Ordovician'. Contemporary interpretations of Wells' statement differ (cf. Whittard 1960; Jones 1938, p. lxxiii; Rast 1969, p. 314).

(ii) Most of the workers of the inter-war years attributed at least some of the volcanics to the Llandeilo on the basis of records of *Glyptogr. teretiusculus*. Many contemporary workers, however, question the stratigraphical significance of this species and, partly on the basis of unpublished field mapping, Rast (1969, p. 318) has attributed much of the volcanics to the Costonian and postulated a substantial break between the Llanvirn and the Caradoc. A little known reference, by Wells (*in* Cox 1925, p. 593) tended to confirm this. He referred to the probability that the volcanic group lay entirely within the *Nemagr. gracilis* Zone as faunas attributed to the Zone (but without the index fossil) occur below the volcanic rocks at Rhobell Fawr (Wells 1925, p. 506—identifications presumably by Elles) and above them at Arenig Fawr (Elles 1922a, pp. 144-5). Addison's reinterpretation of the range of *Nemagr. gracilis*, however, may well support earlier suggestions of the presence of at least some Llandeilo.

(iii) The stratigraphical range of the distinctive Nod Glas—first named from this area (Pugh 1923, pp. 517–8) is not known and its relationship to the Cymerig Limestone not clear. Cave (1965) correlates the Nod Glas with the Gwern y brain Group of Welshpool which contains graptolites attributed to the *Dicranogr. clingani* Zone and a shelly fauna indicative of the *Onnia gracilis* Zone (thus confirming Shackleton's discovery—*in* Whittington 1938, p. 455). He suggests that the deposit represents the Onnian Stage only and that there is, in consequence, a non-sequence below the Nod Glas. The alternative suggestion—that the deposits spanned a much longer period and that there is no non-sequence—may, however, still apply in the type area.

(iv) The remapping of the upper Ashgillian of the area between Aberllefenni and Machynlleth (James & James 1969) has revealed extremely rapid variations in thickness and facies. These may be related to penecontemporaneous movements along a major fault believed to be associated with the Bala fault and a non-sequence near Aberllefenni is tentatively equated with the one at Bala. In neither area, however, is there faunal evidence for the extent of the break. 'Spectacular' variations in thickness attributed to movements along the Bala Fault have also been described in the basement beds of the Arenig (Rast 1969, pp. 313–4).

(E) BERWYN HILLS (Column W 7)

The core of the so-called 'dome' is formed of Ordovician rocks which are arranged in simple stratigraphical sequence along the northern and western limbs, but in much more complicated sequence along the southern overturned limb. The area has not been systematically mapped since the work of the Geological Survey during and immediately after the First World War (Wedd *et al.* 1927, 1929; and also Wills & Smith 1922) but comparatively recent work on restricted aspects has highlighted a number of points. For example:

(i) The three inliers of Llandeilo rocks—the only undisputed rocks of this age in North Wales—have been shown, on the basis of the abundance of *Marrolithus favus* (Salter), to be almost entirely of Upper Llandeilo age (MacGregor 1958, 1961, 1963). The lowest exposed sediments suggest, on lithological grounds, a Llanvirn age (Wedd *et al.* 1929, pp. 19, 33).

(ii) No diagnostic Costonian fossils have been found but Soudleyan faunas have been recorded by Brenchley (1964, p. 44) and G. H. Jones (1956).

(iii) The Longvillian horizon below the Cymerig Limestone (with *Chasmops cambrensis* Whittington) is recognised all around the Berwyn dome (Whittington 1962–68, p. 112). The highes thorizon in the Lower Bala Group, that yielding *Estoniops alifrons*, is recorded (as '*Pterygometopus jukesi*') from the south-western Berwyns (King 1923, p. 491; Wedd *et al.* 1929, p. 44) and the Meifod district (King 1928, p. 698), but not from either the Glyn Ceiriog or Llansantffraid districts. Cave's suggestion (1965, p. 281) that everywhere in North Wales the youngest strata underlying the Nod Glas yield *E. alifrons* is not yet proven.

(iv) The unconformity between the Caradoc and Ashgill, suspected by Whittington (1938), has now been demonstrated along the western limb of the dome·

In Nant Achlas, near Pennant Melangell in the Tanat valley, there is faunal, stratigraphical and structural evidence of a pre-'*P. parabola*' unconformity (G. H. Jones 1956).

(v) The bulk of the Ashgill is represented by the Ddolhir Beds (Groom & Lake 1908) which locally contain a rich fauna (e.g. Ddolhir Beds s.s. of Wills & Smith 1922, p. 187; fauna listed in Whittington 1962-68, p. 123). Correlation with Northern England is, however, somewhat equivocal. The bulk of the fauna suggests a Rawtheyan age, but the presence of *Diacalymene marginata* Shirley may suggest a Cautleyan age (Ingham 1970, p. 17). Specimens apparently referable to *Stygina latifrons* (Portlock) and which probably originate from the Ddolhir Beds near Cynwyd (*see* Whittington 1950, p. 548; Ingham 1970, p. 17) may also indicate a Cautleyan or even earlier age (cf. Killey Bridge Beds fauna). The Lower Trewylan Beds were, on the basis of the major part of the fauna, correlated by Whittington with the Rawtheyan. The presence of *D. marginata*, however, suggested to Ingham that the fauna could be older. Although part of Ingham's evidence is now known to be in error (Whittington 1962-68, p. 117), Price's preliminary studies (1971) tend to confirm Ingham's suggestion.

(vi) The Cyrn-y-brain Beds which are the equivalents of the Ddolhir Beds in the Mynydd Cricor and Cyrn y brain inliers north of the Llangollen Syncline, contain a large fauna (Wills & Smith 1922, p. 191) which has much in common with that in the Ddolhir Beds. The presence of *Eostropheodonta hirnantensis* (M'Coy) and *Hirnantia sagittifera* in the upper parts indicates the presence of the Hirnantian stage. The very top of the Ordovician sequence is represented by a thin series of grits and local limestones known variously as Glyn Grits, Corwen Grits and Plas-uchaf Grits and overlain conformably (Wills & Smith 1922) by the Fron Frys Slates which mark the base of the Llandovery (*see* Cocks *et al.* 1971—column G).

(F) THE CENTRAL WALES INLIERS
(not included in the chart)

Approximately 1,620 m of alternating shales and grits of upper Ashgill age are exposed in the three structural inliers in north Cardiganshire and west Montgomeryshire. The lowest beds (Nant-y-moch Group)—exposed in the Plynlimmon inlier only (Jones & Pugh 1935, p. 264)—contain a graptolite fauna which is attributed to the *Dicellogr. anceps* Zone (Jones 1909, p. 469). They are overlain by approximately 720 m of unfossiliferous strata which are in turn conformably overlain by the *Glyptogr. persculptus* Zone at the base of the Silurian and in the area that it was first defined (Jones 1909, p. 475 *et seq.*)

Since Jones' pioneer work (1909), there has been virtually no advance in biostratigraphy. Stratigraphical mapping by individuals (Jones & Pugh 1916; O. T. Jones 1922; W. D. V. Jones 1945) and by officers of the Institute of Geological Sciences (Annual Reports: 1967, p. 79; 1968, p. 15), sedimentological and petrographic work (James 1968, 1971a, b; James & James 1969) and descriptions of extensive temporary trench sections (Adams 1963; Anderson 1963) have, however, resulted in the refinement of the lithostratigraphical sequence and modifications to the stratigraphical terminology. It has also resulted in very variable estimates of the thickness of the Nant-y-moch Group which range from 270 to 300 m (Annual

Report, Institute of Geological Sciences 1968, p. 79) to 718 m (Jones & Pugh 1935, p. 264). The sequence in these inliers, which is co-terminal with the Silurian sequence represented in Table I of the main Silurian Correlation Chart (Cocks *et al.* 1971), is not represented as a column in the Ordovician Chart.

(G) TOWY ANTICLINE (Columns W 8 and W 9)

The stratigraphical relationships of the Ordovician strata along the eastern flanks of the Towy anticline, and particularly at two of the areas made famous by Murchison, namely Llandeilo and the Builth-Llandrindod inlier, have been studied in considerable detail: parts of the latter area being probably the most intensively mapped of all the Lower Palaeozoic areas of Wales. The stratigraphy of the predominantly late Ordovician strata exposed on the western flank, on the other hand, has not been studied in anything like the same detail, due largely to the apparent paucity of fossils. The axis of the fold corresponds more or less precisely with the interface between the graptolitic and shelly facies of the Upper Ordovician.

Builth-Llandrindod-Abbey Cwmhir (W 8)

The large inlier which extends from Builth Wells to Llandrindod Wells exposes a sequence of Llanvirn and Llandeilo strata which, at a maximum, is some 3,000 m in thickness. It is of importance historically for three reasons. First, because Murchison (1834, p. 14) originally referred to rocks of Llandeilo age as the 'Builth and Llandeilo Flags', although he later (1835, 1839) renamed them 'Llandeilo Flags'. The many references in the various editions of *Siluria* (1854, 1867, 1872), however, indicate that he continued to believe in the importance of strata of the Llandeilo Series at Builth. Second, because Lapworth made the rocks of Llandeilo age at Builth (with equivalent rocks at Abereiddy) the basis of his Middle Division of the Ordovician (1879a). And third, because Elles (1925, 1940) erected the *Glyptogr. teretiusculus* Zone on the basis of the Builth section.

The area is also noteworthy because it contains a 'mixed' fauna in which trilobites and graptolites predominate with occasional horny brachiopods and rare articulate brachiopods. The trilobites and graptolites are frequently found together in abundance on the same bedding planes throughout much of the sequence and it is for this reason that Elles (1940) cited a trilobite and a graptolite as indexes for three of the Zones. It is only in the highest beds of the *Nemagr. gracilis* Zone that the fauna reverts to a dominantly graptolitic one (Elles 1940, p. 433). Recent work on the trilobites confirms that the fauna, although less restricted than was formerly thought, is still quite distinct from other known Anglo-Welsh assemblages (Hughes 1969, p. 45).

The dominant sediments are argillaceous containing thick intercalated sequences of volcanic and pyroclastic rocks which show striking lateral variations. In contrast to the 'remarkable abundance of rock exposures' (Jones & Pugh 1948, p. 67) of the Builth Volcanic Group, the sedimentary sequences are exposed in isolated outcrops only.

The details in the lower part of the column are based on Elles (1940) but include modifications of the volcanic sequence by Jones & Pugh (1941, 1948, 1949) and with the changes in nomenclature suggested by Hughes (1969, p. 46). Those in the upper part are based on Roberts (1929).

Llanvirn Series. The lower part of the 900 m mudstone-shale sequence of the *Didymogr. bifidus* Zone (Elles' '*Didymograptus bifidus* and *Ogyginus corndensis* Zone') is only sparingly fossiliferous. The uppermost 90–105 m, however, yields a rich fauna of trilobites with some graptolites and brachiopods which Elles (1940, p. 397) assigned to a '*Didymograptus speciosus* and *Cryptolithus gibbosus* Subzone'. (*See, however,* Jones & Pugh 1941, p. 186.)

The sequence attributable to the *Didymogr. murchisoni* Zone, which overlies the 'bifidus' beds unconformably, is made up of shales at the base and top (Hughes' 'Lower *Didymograptus* Shales'· and 'Upper *Didymograptus* Shales') intercalated between which are the variable sequences of igneous and pyroclastic rocks of the Builth Volcanic Group and the Newmead Group. The stratigraphical picture is complicated, however, by an unconformity between the two volcanic groups and the absence of the great bulk of the volcanics from the northern part of the inlier, while in the south the upper part of the volcanic sequence, together with the overlying shales, are overlapped by shales containing a *Glyptogr. teretiusculus* fauna.

The fauna of the Zone consists basically of graptolites which are exceedingly abundant at some localities. The zone fossil is present in both upper and lower sequences. Trilobites are generally extremely rare throughout the Zone, while the fauna of the Newmead Group reflects the different facies because it is made up almost exclusively of articulate brachiopods and sponge remains.

Llandeilo Series. Some 600 m of shales and mudstones with an occasional thin limestone and calcareous ash band overlie the *murchisoni* Zone, and because of the complete disappearance of all the tuning-fork graptolites, there appears to be a definite palaeontological breàk (Elles 1940). The incoming fauna is distinctive, containing *Glyptogr. teretiusculus* (Hisinger), *Glyptogr. putillus* (Hall), *Amplexogr. perexcavatus* Lapworth, *Climacogr. antiquus* Lapworth, etc. (Elles 1940, p. 411), and prompted Elles & Wood (1901-18) to introduce the Zone of *Glyptogr. teretiusculus*. Elles later (1925) provided a formal definition and in describing the Zone at Builth (1940, p. 406) referred to it as 'the zone of *Glyptograptus teretiusculus* and *Ogygiocaris buchi*'.

On the basis of the trilobite fauna—which is more abundant than the graptolite one—work in progress by R. Addison, C. P. Hughes and J. K. Ingham (which includes a full reappraisal of the trinucleids of the whole of the region), suggests that the sediments are more or less equivalent to the Lower Llandeilo of Llandeilo.

The sediments containing the *Glyptogr. teretiusculus* fauna grade upward into approximately 300 m of dark indurated shales and mudstones containing a fauna which Elles attributed to the 'zone of *Nemagraptus gracilis* and *Ogygiocaris buchi*'. The passage is so gradual that Elles (1940, pp. 413–5) found difficulty in positioning the boundary even in the almost continuous sections of the river Wye at Pen-ddol Rocks. The presence of *Ogygiocarella debuchii angustissima* and the extension of the range of *Nemagr. gracilis* suggests (Dr. C. P. Hughes, personal communication) that the beds are late Llandeilo and not early Caradoc in age.

Caradoc Series. The oldest post-*gracilis* Caradoc strata so far discovered in the area occur on the western flank of the Towy anticline in the Abbey Cwmhir area (Roberts 1929). The establishment of a detailed faunal succession has not proved possible because of the poor state of preservation of the graptolite faunas. The 240–300 m of soft coal-black shales (Carmel Group) are, however, considered to be equivalent to the *Dicranograptus* Shales of South Wales (Strahan *et al.* 1909, 1914), the lowest beds apparently being attributable to the Zone of *Diplogr. multidens* and the overlying strata to that of *Orthogr. vulgatus.*

Ashgill Series. The presence of a break between the Caradoc (Carmel Group) and the Ashgill (Camlo Hill Group) is suggested by the absence of part at least of the Zone of *Orthogr. truncatus.*

The change in facies across the axis of the Towy anticline is accentuated near Abbey Cwmhir because of the presence of the Carmel thrust. East of the fault the fauna is entirely shelly but it has not been possible to erect any sequence. West of the fault the bulk of the succession—aparently 900 m thick—is unfossiliferous, but some graptolitic and shelly assemblages do occur. The only diagnostic assemblage, however, includes *Eostropheodonta hirnantensis* (M'Coy) and is from massive grits and sandy sandstones in the uppermost part of the sequence. It suggests a Hirnantian age, although the relationship of the strata to those of the *Glyptogr. persulptus* Zone is not entirely clear (Roberts 1929, p. 660). The sequence west of the thrust has, however, been subdivided into three lithostratigraphic units and the striking lithological comparisons with the Ashgill sequences in mid-Wales suggest (Roberts 1929, p. 670) that the *Dicellogr. anceps* Zone is represented.

Garth, Llandovery, Abergwesyn (not included in the chart)
In the areas bordering the Towy anticline from Abbey Cwmhir southwards to Llangadog and Llandeilo, the study of the Upper Ordovician sediments has been neglected. A number of studies over the last half century, however, contain some biostratigraphic information which can be summarised thus:

(i) Sediments of undoubted Caradoc age are known to occur in two areas: Mudstones with gritty and ashy bands overlying volcanic rocks in the Llanwrtyd area (Stamp & Wooldridge 1923, p. 21) yield a basal Caradoc fauna, probably of Costonian age; and black shales exposed on the western flank of the Llanwrtyd volcanic rocks and at the Sugar Loaf contain graptolites attributed to the *Dicranogr. clingani* Zone (Davies 1933, p. 177; Jones 1949, p. 47). The shelly fauna mentioned by Davies (1926, p. 439) from the immediately overlying strata has not been studied. Roberts (1927, p. 294) notes the similarity of the black shales of Baxter's Bank to the Nod Glas or the 'higher beds of the Ceiswyn Group' of North Wales. The relationship of these Caradoc strata to the Ashgill is not known.

(ii) The Ashgill strata west of the anticline reach a maximum thickness of 2,100 m in the Drygarn-Abergwesyn area (Davies 1926, p. 439) and contain a number of horizons of coarse penecontemporaneous conglomerates (Davies & Platt 1933). Fossils are scarce but the lithological sequence is so like that in the Plynlimmon inlier that Davies ascribed the strata to the *Dicellogr. anceps* Zone. On the eastern limb of the fold forms such as *Sampo ruralis* (Reed) and *Sowerbyella* cf. *sladensis* Jones recorded by Jones from the shelly facies of the Ashgill in the Llandovery district may indicate a Cautleyan age.

(iii) The Ordovician–Silurian boundary has been mapped over many miles on both flanks of the anticline especially on lithological differences. In the graptolite sequence west of the anticline the contact is apparently conformable, the overlying strata belonging to the *Glyptogr. persculptus* Zone (Davies 1926, p. 439; 1928, p. 161). On the eastern flank, however, where the Ashgill strata belong to the shelly facies, the contact is unconformable in some if not all areas. In the southern part of the Llandovery area—part of which has recently been 'designated' as the type area of the Llandovery Series (Cocks *et al.* 1970—the work of Jones (1925) and M. A. Woollands (ms.) suggests that the unconformity is not great. In the northern part of the Llandovery area the relations are somewhat uncertain (Jones 1949, p. 48). In the Garth area to the north-east, Andrew (1925, pp. 391–2) reported forms which are usually associated with the basal zone of the Llandovery (*Glyptogr.* cf. *persculptus* and *Mesogr.* cf. *modestus parvulus*) in beds 6 m stratigraphically below the grit which he had taken as the base of the Llandovery Series. If this occurrence is correct it places the base-line of the Llandovery in the middle of a uniform series of mudstones which otherwise contain genera of 'Bala' age.

Llandeilo (W 9)

The critical section at Llandeilo consists of more than 914 m of exclusively shelly facies of late Llanvirn and Llandeilo age intercalated between graptolitic shale units of the lower Llanvirn and Caradoc Series. Since the definitive studies by Murchison (1835, 1839), this type area has been remapped as part of the regional revision by the Geological Survey in the first decade of this century (Strahan *et al.* 1907) and, more recently, as part of a study of the area between Golden Grove and Llangadog (Williams 1948, 1949, 1953). The column is based on this latter work.

Llanvirn Series. The lowest beds exposed in the area include shales and siltstones which contain a *Didymogr. bifidus* fauna (Williams 1953, p. 179). They are subdivided into three members: poorly fossiliferous shales and rhyolitic ashes, with *Didymogr. artus* Elles & Wood as the dominant graptolite; well-bedded finely laminated richly fossiliferous blue-black shales with *Didymogr. bifidus, Didymogr. artus* and *Protolloydolithus ramsayi* (Hicks); and poorly fossiliferous ashy shales in which the commonest fossil is *Dalmanella prototypa* Williams. These beds grade upward into the Ffairfach Group which is a very diverse suite of fossiliferous rocks comprising rhyolitic ash, quartzite, sandstone, conglomerate and minor siltstone and impure limestone grouped into five members and containing a shelly fauna.

The Group was believed by the officers of the Geological Survey (Cantrill & Thomas *in* Strahan *et al.* 1907, p. 12) to be the natural base of the Llandeilo Flags as defined by Murchison. Williams (1953), however, drew attention to the fact that Murchison, in the mistaken belief that the rocks assigned to the Ffairfach Group were correlatives of his 'Caradoc Sandstone', excluded them from his 'Llandeilo Flags'. Williams maintained that, in spite of this error, the distinction is worth retaining in that it 'acknowledges the lithological and physical break at the base of the Llandeilo Series' and that it 'facilitates unqualified acceptance of both Llandeilo and Llanvirn as defined by Murchison (1839) and Hicks (1881) respectively'. The group appears to be a facies variant of part of the *Didymogr. murchisoni* shales and is accordingly better assigned to the Upper Llanvirn.

As stated in the Introduction and discussed more fully later (p. 36), this correlation is upheld by work on the macrofauna but the microfaunal evidence is somewhat contradictory. The occurrence of abundant specimens of the conodonts *Eoplacognathus lindstroemi* (Hamar) and *Prionodus prevariabilis* Fåhraens from the uppermost part of the Ffairfach Group is taken to indicate the *Eoplacognathus lindstroemi* Subzone which equates with the lower part of the *Glyptogr. teretiusculus* Zone in the Swedish succession (Bergström 1971b, p. 109).

The extent of the break between the Ffairfach Group and the basal Llandeilo beds is difficult to gauge. Evidence in the Llandeilo district (Williams 1953, p. 180) suggests that the hiatus represents an insignificant time gap: the Lower Llandeilo fauna is essentially a continuation of that in the Ffairfach Group and the differences in emphasis were probably controlled ecologically (Williams 1969b, p. 240). A correlation of the successions in the Builth-Llandrindod inlier (W 8) with that at Llandeilo, on the other hand, suggests that the sub-Llandeilo break in the type district is substantial. If the Cwm Amliw Ashes are equated with the rhyolitic lava and ash of the Ffairfach Group (Williams 1969b, p. 242, fig. 122) it seems that only a small part of the 122 m of the Upper *Didymogr. murchisoni* Shale exposed near Llandrindod is represented by the variably developed terminal conglomerate member of the Ffairfach Group which is nowhere more than 12 m. thick.

Llandeilo Series. The strata attributed by Williams to the Llandeilo are about 792 m thick. They consist predominantly of flags and impure limestones, though shales and calcareous sands are also important at certain horizons. The Series is for the most part richly fossiliferous and has yielded an extensive, but almost exclusively shelly fauna. By far the most abundant trilobites are the trinucleids. Their widespread distribution in conjunction with rapid evolutionary changes has made it possible to establish three members characterised by distinctive fossil assemblages and, commonly, by equally distinctive types of sediments. They form the basis of the tripartite division of the Series into Lower, Middle and Upper. The lowest is characterised especially by the presence of *Lloydolithus lloydi* (Salter) and *Marrolithus inflatus* Williams and its varieties; the middle, by species of *Marrolithoides*; and the upper, by *Marrolithus* and particularly *M. favus* (Salter).

The Lower Llandeilo is now known, on the basis of Addison's work at Mydrim and Narberth (see later) and Hughes' confirmatory work at Builth, to be equivalent to the *Glyptogr. teretiusculus* Zone.

Conodont faunas have been described from the Lower and Upper Llandeilo Series of Llandeilo by Rhodes (1953, p. 264) and Bergström (1971b, p. 109) respectively. The upper fauna is not in itself diagnostic but on circumstantial evidence does possibly suggest a *Nemagr. gracilis* association.

Caradoc Series. The Llandeilo Flags are succeeded by calcareous graptolitic shales which yield a fauna which is best correlated with the *Nemagr. gracilis* Zone (Williams 1953, p. 194). The zone fossil has not, however, been found. The only reasonably common shelly form is *Talaeomarrolithus radiatus* (Murchison).

The next fossiliferous horizon is the Crûg Limestone, described by Cantrill (*in* Strahan *et al.* 1907, p. 31) and named by Williams (1953, p. 194). It occurs in faulted lenticles and its relationship to associated strata is uncertain. It is, however, considered, on the basis of a rich macrofauna, to be either Longvillian or Marshbrookian (Williams 1953, p. 195) and, especially on the presence of *Kjaerina geniculata* Bancroft, to be equivalent to the Upper Longville Flags and the Robeston Wathen Limestone (Williams 1953, p. 207); and, on conodont evidence, to be Marshbrookian (Bergström 1964, pp. 45, 60; *see also* Lindström 1959).

Ashgill Series. Another lenticular limestone, the Birdshill Limestone, named by Thomas (*in* Strahan *et al.* 1907, pp. 31, 32), contains a large fauna. Shirley (1936) considered the limestone to be the lateral equivalent of the Sholeshook Limestone; Bergström (1964, pp. 51–3, text fig. 22), maintained that the conodont fauna (the youngest currently known from Welsh sections), indicated a Pusgillian age; and Price (1971, p. 272), on the basis of a study of the trilobite fauna, upheld Bergström's suggestion but ruled out exact correlation with the Sholeshook Limestone.

Two faunas are cited by Williams (1953) from the remainder of the Ashgill strata, which is generally unfossiliferous. Both are of Cautleyan age, the lower comparing closely with the 'Calymene beds' of the Cautley district.

(H) SOUTH WEST WALES (Columns W 10 and W 11)

Carmarthen-St. Clears (W 10) Narberth and Haverfordwest (W 11)

Arenig and Llanvirn Series. The strata described as the '*Peltura punctata* Beds' (Crosfield & Skeat 1896, pp. 525, 531; Thomas *in* Strahan *et al.* 1907, pp. 6, 7; 1909, pp. 5–8) and indicated on the one-inch-to-the-mile Geological Survey maps as Tremadoc in age, were shown by Stubblefield (*in* Smith 1933, p. 374) to be of Arenig age.

The age of some of the volcanic rocks associated with the lowermost part of the sequence is, however, still not known. The Trefgarn Andesites, for example, are in faulted contact with Lingula Flags and conformably overlain by beds placed low down in the *Didymogr. extensus* Zone (Thomas & Cox 1924, p. 526). Their general stratigraphical setting has accordingly been compared with that of the Rhobell Fawr sequence. The revised age of the Skomer Volcanic 'Series' (Thomas 1911; Cantrill *et al.* 1916; Ziegler *et al.* 1969), long considered as Lower Arenig, is discussed in Cocks *et al.* 1971.

The lowest sediments of the Arenig are generally arenaceous with some regional variation in grain size (Jones 1938, p. lxxiii) but yielding a shelly fauna throughout —particularly in the Carmarthen (Cantrill *in* Strahan *et al.* 1909, p. 11 *et seq.*) and Henllan Amgoed (Thomas *in* Strahan *et al.* 1914, pp. 17–27) areas.

The basal arenaceous sequence is everywhere overlain by graptolite-bearing strata representing the upper *extensus* and *hirundo* Zones. In the area represented on the Haverfordwest Sheet (Strahan *et al.* 1909, p. 10) the extent of the latter could not be mapped because of its small thickness (60–90 m), the disturbed nature of the strata and the lack of any detectable lithological differences between it and the underlying Zone. The uppermost beds of the Arenig pass conformably into those of the lower Llanvirn with a slight but significant change in lithology. The problems of mapping this junction have been summarised by Jones (1954).

The complicated tectonics of the whole area between Carmarthen and Haverfordwest, added to the lithological uniformity of the greater part of the graptolitic shales of Arenig and Llanvirn age and the thinly distributed graptolites, has meant that no detailed successions have been erected.

The *bifidus* and *murchisoni* Zones have both been recognised (Strahan *et al.* 1907, 1909, 1914). The former represents by far the thickest sequence of strata, but apart from the recognition of a *Didymogr. acutidens* horizon (Thomas *in* Strahan *et al.* 1914, p. 24) no detailed succession has been worked out. The uppermost beds at Lampeter Velfrey and Wolfsdale (Thomas *in* Strahan *et al.* 1914, pp. 25, 30) yield a mixed fauna of graptolites and trilobites yielding '*Trinucleus* cf. *lloydi*'. The fauna at Scolton, Haverfordwest is given by Whittington (1952, p. 8). The *murchisoni* Zone is absent in the Narberth-Haverfordwest area, being overlapped by the Llandeilo.

Llandeilo Series. The shelly sequence in the type area of the Llandeilo Series changes westward into a predominantly graptolitic sequence. In spite of the facies change, however, the succession at Mydrim accords well with that at Llandeilo. The evidence as regards published sources can be summarised thus: The officers of the Geological Survey (Cantrill & Thomas *in* Strahan *et al.* 1909, p. 39) correlated the pyroclastic horizon immediately overlying the graptolitic shales of the Llanvirn (the *Asaphus* Ash) with the upper ashes of the Ffairfach Grit. The topmost horizons of the conformably overlying Llandeilo Flags have yielded a few specimens of *Glyptogr. teretiusculus* (Hisinger) and the immediately overlying shales of the Hendre Shales yield an abundant graptolite fauna which Toghill (1970a) considered to be typical of the *Glyptogr. teretiusculus* Zone. Morris (*in* Toghill 1970a) assigned the trilobite fauna from the base of the Hendre Shales to the Middle Llandeilo on the basis of specimens identified as *Marrolithoides anomalis*. Evans (1906), found *Nemagr. gracilis* in the Mydrim Limestone and Cantrill & Thomas (*in* Strahan 1909, p. 46) subsequently placed the base of the Caradoc Series at its first occurrence—the base of the Limestone—and emphasised that both *Didymogr. superstes* and *Nemagr. gracilis* are restricted to the limestone. Toghill followed Cantrill & Thomas and vigorously disputed Jones' claim (1956, pp. 326, 327) that *Nemagr. gracilis* occurred in the Hendre Shales.

Current work by Robert Addison which will shortly be submitted for a doctoral thesis at The Queen's University, Belfast, has now established the range of *Nemagr. gracilis* with respect to the shelly sequence of the Llandeilo and basal Caradoc Series and, therefore, necessitates a number of changes in the Mydrim sequence and has obvious implications of general significance to the Correlation Chart. Addison states (personal communication, 1972):

'A re-examination of the trilobites described by Morris (*in* Toghill 1970[a]) from the base of the Hendre Shales failed to confirm the occurrence of *Marrolithoides anomalis*. The presence of *Lloydolithus lloydi* and '*Cryptolithus*' sp. in the Llandeilo Flags and the lower part of the Hendre Shales, however, indicates a Lower Llandeilo age. Furthermore, specimens of a *Marrolithus inflatus* s.l. indicate either an Upper Llanvirn or Lower Llandeilo age.'

'A number of specimens of *Nemagraptus gracilis* have been found at horizons well below the level of the Mydrim Limestone in the lane sections near Ty Newydd Farm (Grid Ref. 3618,2208 and 3564,2195) and the identifications have been confirmed by Dr. Toghill. The graptolites occur in association with *Lloydolithus lloydi*.

'The occurrence of *Lloydolithus* is considered significant as Williams (1948, p. 87; 1953, pp. 190–2) showed that the genus was restricted to the lower Llandeilo of the type area. Evidence from the Shelve area was interpreted differently by Whittard (1955-67) who placed the Rorrington Beds with their fauna of *Nemagraptus gracilis* in the Caradoc leaving the Meadowtown Beds as the only representatives of the Llandeilo Series in the Welsh Borders. By this decision the range of *Lloydolithus* which occurs throughout the Meadowtown Beds was extended to the top of the "Llandeilo" of the Shelve area. Similarly the occurrence in the Rorrington Beds of *Marrolithoides arcuatus*, which Whittard stated to be almost identical to *M. simplex*, extended the apparent range of this middle Llandeilo zonal index into the "Caradoc" of Shelve. These apparently conflicting ranges in Shelve and Llandeilo fall into line, however, if the base of the *gracilis* Zone is taken at some level just below the top of the lower Llandeilo, as might be suggested from the Mydrim faunal succession. Thus the Meadowtown Beds are correlated with the lower Llandeilo and the Rorrington Beds with the middle and upper Llandeilo and partly with the lower Costonian.'

The relationship of the shelly sequences of the upper Llandeilo to that of the lowermost Caradoc is revealed in the Narberth area, also studied by Addison. He states:

'The Llandeilo Flags exposed at Lampeter Velfrey and Llandewi Velfrey, east of Narberth and hitherto placed in the lower Llandeilo (Cantrill, Thomas & Jones *in* Strahan *et al.* 1914, pp. 31-6); and, with qualifications, Spjeldnaes 1963, p. 262) are, on the basis of the shelly faunas, partly of upper Llandeilo and partly of basal Caradoc age.'

'The evidence was obtained from three sections which together provide the full succession. The lower part of the succession consists of 60 m of richly fossiliferous flags and limestones which contain *Marrolithus favus* and succeed shales containing *Didymograptus bifidus* although the contact is not seen. An overlying sequence of limestones 70 m thick in Bryn Banc quarry north-east of Llan-mill yields undescribed marrolithinid faunas distinctly different from any faunas known from the Llandeilo beds of Llandeilo but resembling in part *Marrolithus ventriculatus* Whittard from the Spy Wood Grit of Shropshire. The age of these beds is not clear as the fauna which includes *Brongniartella, Flexicalymene, Decoroproetus, Metopolichas, Heterorthina, Oxoplecia, Platystrophia, Multispinula, Mcewanella, Glyptorthis* and *Dalmanella*, is not truly diagnostic. The third section, north of Stoneyford, largely overlaps that seen in Bryn Banc quarry but shows a further 15 m of the fossiliferous limestones grading up through 12 m of largely unfossiliferous flags into black *Dicranograptus* shales yielding *Nemagraptus gracilis* (Strahan *et al.* 1914, p. 41). The limestones near their top contain *Costonia elegans* Dean as well as *Dalmanella, Glyptorthis, Heterorthina, Horderleyella, Platystrophia, Skenidioides, Sowerbyella, Macrocoelia, Flexicalymene* and *Primaspis* which indicate a Costonian age and a conodont fauna of *"gracilis"* zone age (Bergström, personal communication).'

Caradoc and Ashgill Series. The greater part of the Caradoc is represented by graptolitic shales. In the Mydrim Shales, for example, the *gracilis, multidens, clingani* and possibly the *linearis* Zones are represented (Cantrill & Thomas *in* Strahan *et al.* 1909, p. 46; Toghill 1970a). In the earlier literature the *Diplogr. multidens* Zone is represented by three Zones—*Diplogr. multidens, Orthogr. calcaratus vulgatus* and *Orthogr. truncatus* (Strahan *et al.* 1909, 1914, p. 40).

The graptolitic sequence is generally succeeded by fossiliferous limestone developments of very variable lithology (including the Crûg and Birdshill Limestones already mentioned) and these by the Redhill and Slade Mudstones. Marr & Roberts (1885) equated the Robeston Wathen and Sholeshook limestones of the Haverfordwest district with 'parts at any rate' of the 'Bala' (Cymerig) and Rhiwlas Limestones of Bala; and Marr (1907), in defining his Ashgill Series, placed the Robeston Wathen and Sholeshook limestones at the top of the Caradoc and the base of the Ashgill respectively and remarked that as far as Britain is concerned 'we find the most satisfactory development of the Ashgillian Beds' in South Wales.

The officers of the Geological Survey (Strahan *et al.* 1907, 1909, 1914) found difficulty in distinguishing the various limestone developments and referred to them collectively as 'Bala Limestones'. In the almost complete absence of work during the succeeding half century, one of the limestones (Sholeshook) has, with reference to work in other areas (e.g. Marr 1913; King & Williams 1948), become widely recognised as Middle Ashgill in age. More recently Price (1971), on the basis of a detailed study of the trilobites of the Sholeshook Limestone, and by comparison with Ingham's work at Cautley (1966, 1970), has suggested that the limestone ranges from Zone 1 to probably the upper part of Zone 3 of the Cautleyan Stage (*see also* Cave 1960).

If this suggestion is correct, several elements of the *Phillipsinella parabola—Staurocephalus* fauna occur in strata which are of early Ashgill age and this calls into question the stratigraphical significance of the *S. clavifrons* Zone and also the attribution of the Trewylan Beds of the Berwyn Hills to the Middle Ashgill (*see* p. 28).

Regarding the uppermost part of the Ordovician succession, the St. Martin's Cemetery Beds of the Haverfordwest sequence have commonly been accepted as lowermost Silurian in age following the work of Strahan *et al.* (1914, pp. 77 *et seq.*) and Jones (1925, p. 354). The presence of a *Hirnantia* type fauna in these beds, however, indicates a late Ordovician age (Ingham & Wright 1970) and this is confirmed by the discovery of *Tretaspis* sp. (Cocks 1968) in the overlying Cartlett Beds which therefore span the Ordovician-Silurian boundary (*see also* Cocks *et al.* 1971, p. 107).

Ramsey Island, Llanvirn and Abereiddy Bay
The Ordovician rocks of north-west Pembrokeshire and southern Cardiganshire are exposed in a series of coastal sections and quarries on Ramsey Island and on the mainland from Whitesand Bay to beyond Cardigan. The southern exposures are noteworthy because of the well-developed shelly faunas of the Arenig on Ramsey Island and historically because the Llanvirn was erected by Hicks (1881) on the basis of the succession from Llanvirn-y-frân Farm to Abereiddy Bay. The northern exposures have not been studied with anything like the same intensity.

Arenig Series. The base of the Series—both on Ramsey Island (Pringle 1930, p. 9; Bates 1969b, p. 4) and on the mainland (Cox 1916, p. 283; Jones 1940; Evans 1948)—rests disconformably on Lingula Flags. As in North Wales *Bolopora undosa* has been found in the basal sediments (Cox *et al.* 1930, p. 422; Bates 1969a, p. 156).

The lower part of the *extensus* Zone is represented by arenaceous sediments on Ramsey Island (Bates 1969b, p. 4) and at a number of localities on the mainland. The palaeontological differences listed by Pringle (1930) cannot, however, be used to separate the apparent equivalents of the Abercastle and Porth Gain Beds on the island (Bates 1969b, p. 4) and the term Ogof Hên Formation has been proposed instead.

The shelly fauna differs from the predominantly brachiopod fauna of Anglesey (Bates 1969a, p. 157) in being relatively poor in brachiopods and in that the latter has distinct Baltic affinities, whereas the former has no Baltic elements. The trilobite *Megaspidella murchisoniae* (Murchison) (Bates 1969b; Whittington 1966) described from these strata of *extensus* Zone age is also known from North Wales and the Welsh Borderland. It appears to be a useful indicator of beds of this age.

Llanvirn and Llandeilo Series. In the type section—from Llanvirn-y-frân Farm northwards to Abereiddy Bay—the *Didymogr. bifidus* Shales succeed the *Tetragraptus* Shales 'without any break, and in fact without any very perceptible lithological change'. So much so that Cox maintained that in the absence of fossils 'it becomes almost impossible to draw any exact boundary-line between the two groups' (1916, p. 293). Flaggy ashes near the base are overlain by the more typical slates which contain the rich faunas collected and described by Hicks (1875), Hopkinson & Lapworth (1875) and Elles (1904).

The conformable junction with the overlying *Didymogr. murchisoni* Zone is marked by a 45 m ash (the *Didymogr. murchisoni* Ash) which in turn is overlain by dark shales containing a *murchisoni* fauna.

These shales were believed by Cox (1916, p. 304) to grade upwards into the *Dicranograptus* Shales. O. T. Jones (1940), however, demonstrated that the original relationships have been destroyed by faulting; nevertheless he suggested (partly on the basis of the relationship in other Pembrokeshire sections) that the junction represented a non-sequence if not an unconformity.

The poorly fossiliferous black shales which occupy the entire central part of Abereiddy Bay have not been differentiated and were allocated by Cox to the *Dicranograptus* Shales. The structurally overlying Castell Limestone was considered to be stratigraphically higher and equivalent to the Mydrim Limestone.

The ideas regarding the age and the sequence of the beds in the outcrops north of the Bay—believed by Cox to be a depleted repetition by faulting of the succession in Abereiddy Bay—have been radically changed within the last two or three years. Waltham (1971), on the basis of cleavage-bedding relationships and the detailed morphology of an unconformity between the Castell Limestone and the *Dicranograptus* Shales, stated that the sequence was inverted, forming the northern limb of an overfolded syncline cut along its axis by a major fault and that, in consequence, the Castell Limestone was below rather than above the *Dicranograptus* Shales. Black *et al.* (1972) have since confirmed the presence of an overfold on faunal as well as structural grounds. They also show that the succession in the northern, inverted limb is much more complete than hitherto believed and they question the existence of the thrust inferred by Cox and Waltham. The occurrence of the *bifidus* Zone is confirmed; a fauna of *murchisoni* Zone age found; the 'trilobite-bed' referred by Hicks (1875) to the Llandeilo and transferred by Cox to the lower Llanvirn is re-allocated, on the occurrence of *Cnemidopyge* cf. *parva* Hughes and *Platycalymene* cf. *tasgarensis simulata* Hughes, to the lower Llandeilo, being thus in accord with Bergström's suggestion (1964) based on a very close correlation of the conodonts with those in the Llandeilo Limestone of Llandeilo; and a trilobite fauna from shales stratigraphically overlying the Castell Limestone (Cox's *Dicranograptus* Shales) allocated on the occurence of *Trincucleus fimbriatus* Murchison and *Telaeomarrolithus* cf. *intermedius* Hughes to the *Nemagr. gracilis* Zone, and suggested as the lateral equivalent of the Mydrim Limestone.

The sequence described by Black *et al.* confirms the suggestion by O. T. Jones regarding the nature of the Llanvirn-Llandeilo contact in the type section.

Caradoc and Ashgill Series. The *Dicranograptus* Shale facies undergoes an abrupt transition into coarse feldspathic greywackes interbedded with black shales when traced into the area around Newport (Pembrokeshire) and Cardigan. Although these arenaceous rocks were correctly placed in the Caradoc by Keeping (1881 and particularly 1882) the area has been persistently represented as Silurian on the majority of maps (*see* Challinor 1951 for details). It is, for example, misrepresented in the South Wales Regional Handbook (George 1970).

The overlying Ashgill sediments are poorly known. In the Llangranog area green and grey mudstones, containing a rich graptolite fauna preserved in full relief and attributed to the *Dicellogr. anceps* Zone (Hendricks 1926, p. 123) are overlain by an unfossiliferous group of sandstones and mudstones exhibiting rapid changes of thickness (143–802 m) and lithology (Anketell 1963). This in turn is overlain conformably by shales and mudstones containing a basal Llandovery graptolite fauna.

4. The Welsh Borderland

By WILLIAM THORNTON DEAN

EXCLUDING the Welshpool area and part of the Breidden Hills, the greater part of the Ordovician rocks of the Welsh Borderland occurs in Shropshire, where they comprise two distinct successions: the Shelve Inlier, ranging from Arenig to Caradoc Series; and the Caradoc District and adjacent areas, comprising strata of the Caradoc Series only. The Church Stretton Fault System or Disturbance, although often supposed to be important in separating different facies in the Ordovician, is, in fact, traversed by certain Caradoc strata and faunas (see later), and the major structural feature separating the Shelve Ordovician rocks from those farther east is the Linley–Pontesford Fault.

(A) SOUTH SHROPSHIRE (Columns WB 1 and WB 2)

The first and more southerly of these includes the type section of the Caradoc Series in the Onny Valley and is separated from the second by the Pre-Cambrian rocks of Hope Bowdler Hill. Differences between WB 1 and WB 2 are relatively minor, but diachronism and lateral facies changes are common place throughout the whole district, which forms part of the eastern margin of the Welsh Lower Palaeozoic Geosyncline. The rocks are predominantly of calcareous, shallow-water type and comprise sandstones, siltstones, mudstones and shales which contain abundant shelly fossils. The area was the scene of Bancroft's original researches (1933) on the Stages and Zones of the Caradoc Series, whilst recent detailed accounts of the geology are to be found in papers by Dean (1958, 1964) and the Geological Survey's Church Stretton Memoir (Greig *et al.* 1968).

The basal arenaceous beds of the Caradoc Series are diachronous and rest with marked unconformity on rocks of Pre-Cambrian, Cambrian or Tremadoc age. The lower half of the Costonian Stage of WB 2 contains the zonal graptolite *Nemagr. gracilis*, but is correlated with the middle third of the Costonian of WB 1.

According to Jaanusson & Strachan (1954) the *Climacogr. peltifer* and *Climacogr. wilsoni* Zones of the Southern Uplands are replaced in the Anglo-Welsh area by a single *Diplogr. multidens* Zone. The graptolitic sequence for WB 1 and WB 2 has recently been modified slightly, and the *multidens* Zone is now considered to commence with the highest subdivision of the Costonian Stage (Dean 1967), by analogy with the Spy Wood Grit of the Shelve Inlier; the latter is of *multidens* age and both it and the topmost type Costonian contain *Costonia ultima* (Bancroft). The *multidens* Zone extends upwards to include at least the lower Soudleyan Stage, and perhaps also the upper Soudleyan, though graptolitic evidence is uncommon. At the top of the type Caradoc succession the graptolitic evidence, though again not abundant, seems reliable and shows that the Actonian Stage and most of, if not all, the Onnian Stage, which are together correlated with Stage 4bδ of Southern Norway, belong to the *Dicranogr. clingani* Zone (*see also* WB 6, Welshpool District). Undoubted post-*clingani* strata are either absent or obscured by unconformably overlying Llandovery strata. There is no unequivocal evidence in South Shropshire for the age of the Longvillian and Marshbrookian Stages in terms of graptolite zones, and the boundary between the *multidens* and *clingani* Zones cannot be drawn with certainty.

(B) PONTESFORD (Column WB 3)

This geographically and stratigraphically restricted outcrop, situated by the west side of the Longmynd, has recently been redescribed by Dean & Dineley (1961). The area lies east of the Linley-Pontesford Fault but some distance west of the Church Stretton Fault, and the succession is closely related to that of the Caradoc District. A poorly-sorted basal conglomerate resting on Pre-Cambrian rocks is followed by the lower Pontesford Shales, which resemble lithologically the Harnage Shales of Harnage Brook (in area WB 2), and the southernmost outcrops in Pontesford Brook contain abundant graptolites, including the zonal species *Diplogr. multidens* Elles & Wood, for which this is the type area. The succeeding higher Pontesford Shales are grey-green mudstones with trilobites indicating the Glenburrell Beds, upper Soudleyan Stage, of the Onny Valley (area WB 1), together with graptolites of the *multidens* Zone. The uppermost part of the section is obscured by unconformable Carboniferous strata and Quaternary deposits.

(C) SHELVE INLIER (Column WB 4)

The stratigraphy of what is perhaps the best-developed Ordovician succession in the Anglo-Welsh area was first elucidated by Lapworth & Watts (1894), but our present detailed knowledge is largely the result of extensive work by Whittard (1931, 1952, 1955-67) and the thicknesses shown in the accompanying tables are based on figures published by him. The succession which is remarkably complete and ranges from lower Arenig Series to the lower Soudleyan Stage of the Caradoc Series, is largely argillaceous but siltstones and volcanic rocks are well represented.

The basal strata, the Stiperstones Quartzite, rest unconformably on the Habberley Shales, of Tremadoc age. Although almost unfossiliferous they have yielded rare *Neseuretus*, a trilobite genus unknown below the Arenig Series and found abundantly in the overlying Mytton Flags. The latter comprise siltstones and shales in which mixed shelly and graptolitic assemblages occur frequently, and the distribution of graptolite zones is reasonably well established. Trilobites in this and higher parts of the succession are stratigraphically important and

Series	Stage
Ashgill	Hirnantian
	Rawtheyan
	Cautleyan
	Pusgillian
Caradoc	Onnian
	Actonian
	Marshbrookian
	Longvillian
	Soudleyan
	Harnagian
	Costonian
Llandeilo	upper
	middle
	lower
Llanvirn	upper
	lower
Arenig	upper
	lower

FIG. 7—Chart sho
Borderland.

exhibit affinities with the Mediterranean region and Bohemia. The boundary between Weston Beds and Betton Beds cannot be distinguished lithologically, and has been drawn on faunal grounds. Similarly the subdivisions of Betton Beds, Meadowtown Beds and Rorrington Beds grade successively into one another, but the trilobite faunas are distinctive (*see* Whittard 1955-67, pp. 265-306). The Meadowtown Beds are equated with the *Glyptogr. teretiusculus* Zone but there is as yet no record of the zonal graptolite, though *Diplogr. foliaceus* (Murchison), for which this is the type area, is regarded as of zonal significance. The Rorrington Beds were at one time stated to belong to both the *Nemagr. gracilis* and the succeeding *Diplogr. multidens* Zones (Whittard 1955-67, p. 5), though a later publication (Whittard 1960, p. 237) assigned them merely to the former Zone (*see also* Addison *in* Bassett herein). Further refinement of this part of the sequence awaits the description of the graptolites. The succeeding Spy Wood Grit, though largely arenaceous, contains graptolites of the *multidens* Zone as well as the trinucleid *Costonia ultima* (Bancroft), characteristic of the uppermost third of the Costonian Stage in the Caradoc district (*see* WB 1). Diagnostic Harnagian Stage trilobites of WB 1, WB 2 and the Welshpool district (WB 6) have not been found in the Shelve Inlier but the higher part of the Aldress Shales contains the lower Soudleyan trinucleid assemblage of *Broeggerolithus broeggeri* (Bancroft) and *Salterolithus caractaci* (Murchison), both of which range upwards through most of the remainder of the succession into the Whittery Shales, the highest subdivision. There is no faunal evidence for the upper Soudleyan or higher Stages of the Caradoc Series in the Inlier, and by analogy with WB 1 and WB 2 the succession should not extend higher than the *multidens* Zone.

(D) BREIDDEN HILLS (Column WB 5)

The stratigraphy of the Ordovician rocks of the Breidden Hills, on the border between Montgomeryshire and Shropshire, was described by Watts (1885, 1925) aud Wedd (1932), and has since been reviewed briefly by Whittard (1952). The approximate thicknesses shown in the accompanying tables are derived from the 1885 account by Watts, as no more recent figures have been published. The lowest strata visible are black shales with *Nemagraptus*, so that presumably at least part of the *gracilis* Zone is present, though the relationship of the shales to underlying rocks is obscured by the alluvium of the Severn Valley. Trilobites and brachiopods from the highest part of the succession indicate no strata younger than Soudleyan Stage (Mr J. Standring, personal communication), but the precise distribution of Harnagian strata is not known. Stubblefield (*in* Wedd 1932, p. 52) demonstrated that all the succession above the Black Grit, the precise age of which is unknown, belongs to the *multidens* Zone, and this is in agreement with what is known of the position of the Harnagian and Soudleyan Stages in WB 1 and WB 2. Whittard (1952, p. 161) noted that the Ordovician rocks of this area may have greater affinities with the Shelve Inlier than with other areas of Shropshire, and as the succession lies west of the Linley-Pontesford Fault, his surmise may well be true. If this should be the case, then the Breidden volcanics may be related to those of the Hagley and Whittery Groups in the Shelve Inlier, also of Soudleyan age.

(E) WELSHPOOL (Column WB 6)

The Ordovician succession here was established first by Wade (1911), but the first detailed correlation was that attempted by Bancroft (1933), whose work has since been elaborated by Cave (1957, 1965). The lowest strata, the Trilobite Dingle Shales, belong to the Harnagian Stage of the Caradoc Series, and the presence of *Salterolithus caractaci* (Murchison) indicates affinities with both the Shelve Inlier (WB 4) and the Onny Valley (WB 1). By analogy with these areas, the *multidens* Zone is represented. According to Cave (1965, p. 287) the *Onnia gracilis* Zone of the Onnian Stage, as represented by the Pen-y-Garnedd (=Gwern-y-brain) Shales, contains graptolites that are no younger than *clingani* Zone.

5 Eastern England

By ISLES STRACHAN

S UBSURFACE Ordovician rocks are known at a few places in the east and south-east of England. The striking absence of any Ordovician in the Midlands above the widespread Cambrian may be partly original but is certainly partly erosional although no trace of fossiliferous Ordovician beds is known from Carboniferous and later conglomerates which yield Silurian pebbles of apparently local origin. The Ordovician pebbles in the Bunter Pebble Beds are clearly far travelled.

In older boreholes, little basement was penetrated and resulting faunas are sparse. The Culford bore, about 5 miles north-north-west of Bury St. Edmunds, went into unfossiliferous material considered as pre-Coal Measures (Whitaker & Jukes-Browne 1894) but does not seem to have been reassessed. Other occurrences have been shown to be Silurian (Cocks *et al.* 1971, pp. 124–5). However re-examination of the Bobbing (Kent) material has led to the claim that the macrofossils indicate an Upper Ordovician age, a claim upheld by the palyno-logical evidence which supports a Caradoc age (Lister *et al.* 1969).

Two more recent boreholes have provided good evidence of Lower Ordovician rocks at depth.

At *Great Paxton*, about 3 miles north-east of St. Neots (TL 2088 6389), Ordovician (Llanvirn) underlies Lower Lias and consists of steeply dipping (60°–75°) mud-stones and thin silt bands. About 27 m of beds stratigraphically were penetrated and yielded an abundant fauna including *Didymogr. murchisoni* (Beck), *Didymogr. artus* Elles & Wood, *Didymogr. acutus* Ekström, *?Aulograptus* sp., *Glyptogr.* cf. *euglyphus* (Lapworth), *Climacograptus* spp.; *Geragnostus* sp., trinucleids including *?Protolloydolithus* sp., *Lonchodomas* sp., cyclopygids, *?Ogygiocaris* sp., *Flexicalymene* sp., *Bohemilla* sp., *Triarthus* sp.; brachiopods, molluscs and ostracodes. A preliminary account of the occurrence has been given by Stubblefield (1967) who noted that the affinities of the fauna were with those of Shropshire, South Wales and Brittany.

The second Survey borehole to prove Ordovician, at *Huntingdon* (TL 2369 7143), has also shown Llanvirn beds below Trias. A summary log of the borehole was reported in the Annual Report of the I.G.S. for 1966 and the fauna in that for 1967. The beds, with low dips but constantly crumpled and with tectonic lineations, consist of dark grey silts and mudstones with red and green partings. Fossils are infrequent but include *Didymogr.* cf. *bifidus* (Hall), *Didymogr.* cf. *artus* Elles & Wood, *Glyptogr. dentatus* (Brongniart); with rare *Ectillaenus* sp., *Pricyclopyge sp.* and *?Cornovica* sp., indicating a generally similar age and province to the Great Paxton beds.

I am indebted to Dr. A. W. A. Rushton for some of the above details.

6. The North of England

By JOHN KEITH INGHAM AND ANTHONY DAVID WRIGHT

WITHIN the North of England, the subdivision of the Ordovician rocks is stratigraphical rather than areal, the successive groups being very similar over the entire region. The Lower Ordovician Skiddaw Slate Group is structurally complex, and the stratigraphic sequence controversial; the evidence of the graptolite faunas is as yet insufficient to establish the precise time interval represented by the sequence. The overlying Borrowdale Volcanic Group is essentially subaerial, and although local volcanic sequences are well known, their regional correlation is very problematical, as is any exact dating of the onset and cessation of the lava outpourings.

The Upper Ordovician is represented by the Coniston Limestone Group, a sequence of essentially shallow water facies in which fossils are relatively abundant. Faunal studies and detailed mapping, chiefly in the inliers to the east of the Lake District, have resulted in fairly sound dating and correlations of this part of the succession with sequences both inside and outside the region.

(A) THE LAKE DISTRICT (Columns N1 to N3)

Since the reviews of Lake District geology by Hollingworth (1954) and Mitchell (1956a), much work has been carried out in the area by many workers. Despite this, there is no general agreement on the many problems of the two very complex Lower Ordovician Groups. A full bibliography of the earlier works is given by Smith (1965).

Skiddaw Slate Group. The main outcrop lies almost entirely in our northern Lake District region (N3). Structural complexities make the establishment of any lithological sequence controversial, and the very restricted records of graptolite faunas considerably reduce their value for correlation. Various formational names and sequences have been proposed (Ward 1876; Rastall 1910; Dixon 1925; Eastwood *et al.* 1931; Rose 1954; and Jackson 1961, 1962) while recent works on large areas of the main outcrop are those of Simpson (1967) and Eastwood *et al.* (1968). The sequence of the latter (N3, Cockermouth) shows a gradual decrease of grain size in the clastic succession, with the unfossiliferous Grits Formation at the base. The overlying Slates and Sandstone Formation contains Arenig graptolite faunas of the Zones of *Didymogr. deflexus* and *Didymogr. nitidus* as do Dixon's Loweswater Flags in the Keswick District (Jackson 1961), which again rest on unfossiliferous strata. The succeeding Slates Formation passes up into the Zones of *Isogr. gibberulus, Didymogr. hirundo* and *Didymogr. bifidus* (Eastwood *et al.* 1968, p. 31), although the Mungrisdale locality of Elles (1933) to the east is suggested by Skevington (1970a, p. 405) as being of high *bifidus* age. Eastwood *et al.* record the Skiddaw Slates as being interbedded with lavas in continuous sequence with the Borrowdale Volcanic Group, although the outcrops in question have a faulted relationship with the main outcrop of Borrowdale Volcanics (but see below).

Working over a larger area of the Skiddaw Slate outcrop, Simpson (1967) recognised eight formations totalling over 9000 metres. His structural interpretations result for example in the grits being placed higher in his succession than the Loweswater Flags; and in the junction between the Skiddaw Group and the Borrowdale Group being an angular unconformity rather than the more generally envisaged conformable relationship (Hollingworth 1954; Mitchell 1956a), with the Latterbarrow Sandstone as the lowest member of the Borrowdale Group.

In the eastern Lake District (N2), the Skiddaw Slates are poorly exposed in the Bampton and Ullswater Inliers. The highest fauna yet recorded has come from one section of the Tarn Moor tunnel between Hawes Water and Ullswater; the evidence of the graptolites and the microfauna suggests the *Didymogr. murchisoni* Zone (Wadge *et al.* 1970). The *Didymogr. bifidus* Zone is present in the Ullswater Inlier (Elles 1933), and while the presence of high *bifidus* Zone in the Bampton Inlier has been well substantiated (Skevington 1970a), evidence of the earlier, Arenig, zones is lacking.

Work recently completed (Wadge 1972) in the northernmost part of the eastern region (N3), to the east of Keswick, documents a number of important sections which reveal the unconformable contact between the Skiddaw Slates and the Borrowdale Volcanics in that area. A thick basal conglomerate, thinning westwards, forms the basal unit of the Borrowdale Volcanic Group and the unconformity is shown to be an angular one; the Skiddaw Slates are believed to belong to the *bifidus* Zone. The fact that these Skiddaw Slates have interbedded volcanic rocks older than the main Borrowdale Volcanic sequence throws some light both on the faulted Cockermouth succession (Eastwood *et al.* 1968), referred to above and also the Cross Fell Skiddaw Slate sequence (N4).

In the south-west (N1), an assemblage indicative of high *Didymogr. hirundo* to low *Didymogr. bifidus* age has been recorded from the Greenscoe Inlier (Knipe & Grieve *in* Soper 1970, p. 488); the fossils recorded by Smith (1912, p. 406) and Helm (1970, p. 109) from the Black Combe Inlier are unfortunately not diagnostic but here Helm (*op. cit.*, p. 111) has demonstrated the unconformable relationship between the Skiddaw Slates and the Borrowdale Volcanics.

Borrowdale Volcanic Group. The junction between the Skiddaw Slates and the Borrowdale Volcanics is largely a tectonic one; in some areas, however, there is evidence of a marked unconformity, while in other sections alternating sediments and volcanics have been regarded as indicating a continuous sequence. Opinion on the form of the junction is sharply divided (*see* Soper 1970) although current studies seem to be clarifying the situation.

The volcanics themselves are dominated by andesitic lavas and tuffs, with less abundant rhyolites; but there is considerable variation in both the lavas and the pyroclastics. For details of the local volcanic sequences of the main belt of the Borrowdale Group and their correlation, the works of Mitchell (1929, 1934)

| Arenig | Llanvirn | Llandeilo | Caradoc | Ashgill |

FIG.

Slates lacks diagnostic fossils although Burgess (*in* Helm 1970, p. 136) regards them as Arenig in age. The Ellergill Beds are an argillaceous sequence containing graptolite faunas of *bifidus* age, while Shotton regarded the Milburn Beds as being of high *bifidus* age and, as they consist of shales with graptolites interbedded with ashes and andesitic lavas, as giving an alternation of Skiddaw Slate and Borrowdale volcanic facies at the end of early Llanvirn times. The Milburn Beds are nowhere separated from other beds by a normal junction and, according to Shotton, may be at least partly equivalent to the Borrowdale ashes with andesites which overlie undifferentiated Skiddaw Slates at Keisley and Roman Fell. Recent discussion, however, on the microfossil and graptolite evidence by Lister and Skevington (*in* Skevington 1970a, pp. 442–4) suggests that while both Ellergill and Milburn Beds are of *bifidus* age, the Ellergill beds are the younger. This interpretation implies the existence of volcanic rocks within the Skiddaw Slates which are wholly earlier than the Borrowdale Volcanic Group *sensu stricto*, a situation similar to that described by Wadge (1972) from the eastern Lake District (*see above*).

Two other isolated and distinct groups of Borrowdale Volcanics occur in the inlier. Near Melmerby there occurs a faulted block of amygdaloidal andesite, with a suggested correlation with the Wrengill Andesites of the eastern Lake District (Green 1915, p. 215; Mitchell 1929, p. 20) and a small quantity of porphyritic andesite which petrologically strongly resembles the Eycott Hill Lavas of the northern Lake District (Harker 1891). Elsewhere in the inlier, and attaining its greatest thickness on Dufton Pyke, occurs a group of rhyolites which are correlated with the Upper Rhyolites of the Lake District. The base is not seen, and although Shotton (1935, p. 654) points out that a small area of the rhyolite on Roman Fell is followed in a normal manner by *Corona* Beds, he assumes that this upper junction is unconformable (p. 658). The possibility of a normal junction on Roman Fell is commented on by Dean (1959, p. 190), and Burgess (1965, p. 57) has reported nodular rhyolites overlain by purple 'ashy' siltstones which pass up into purple mudstones with a *Corona* Beds fauna. 'Ashy' suggests that volcanic activity persisted after sedimentation of the *Corona* Beds started in lower Longvillian times, but the evidence as yet cannot be regarded as sufficient to indicate a conformable passage or a stratigraphic break at the top of the Borrowdale Volcanic sequence.

The Caradoc faunas of the overlying sediments are similar to those of the type Caradoc area of Shropshire (Bancroft 1933; Dean 1959, 1962). The term *Corona* Beds (Nicholson & Marr 1891) has been restricted recently by Dean (1959) to the essentially purple and grey mudstones of lower Longvillian age. The overlying Dufton Shales, principally dark grey calcareous mudstones and shales, contain well-documented shelly faunas which extend from upper Longvillian through to the Pusgillian and lowest Cautleyan (Zone 1) Stages of the Ashgill Series. The separate term 'Melmerby Beds' for lower and upper Longvillian shales near Melmerby (Dean 1959, pp. 210–14) has not been used in the chart. The Dufton Shales are overlain unconformably by the Rawtheyan Swindale Limestone which has a fauna similar to that of Zone 6 in the type Ashgill sequence at Cautley (Ingham 1966). The local representative of the Ashgill Shales (mainly Hirnantian but probably including a small thickness of beds of highest Rawtheyan age) is faulted against the Swindale Limestone in Swindale Beck.

The chief correlation problem in the Coniston Limestone Group of the Cross Fell Inlier is that of the Keisley Limestone. Marr (1906, p. 482) correlated it with the Swindale Limestone; Ingham (1966, p. 494) suggested that it could be correlated with an horizon 'fairly low in the Ashgill Series of Cautley' on the basis of Reed's faunal lists (1896, 1897); while Wright (1968, p. 366), by comparison with the Kildare Limestone in Ireland, believed it to be relatively high in the Ashgill, a view evidently supported by Whittington (1962-68, p. 121). As the stratigraphic relationships of the fault-bounded reef limestone are not clear and, as there is the further complication of a facies fauna, any unequivocal dating of the Keisley Limestone must await the revision of Reed's faunal studies.

Although the fauna obtained from thin limestone bands apparently overlying the Keisley Limestone was regarded by Temple (1968) as being of Lower Llandovery age, it does contain elements of the *Hirnantia* fauna and is considered by us to be better interpreted as uppermost Ashgill in age.

To the east of Cross Fell, rhyolitic tuffs of the Borrowdale Volcanic Group together with Skiddaw Slates are reported from the small inlier at Langdon Beck in upper Teesdale (Dakyns 1877; Johnson 1961), and it seems certain that these Lower Ordovician rocks extend over a considerable area of northern England beneath the Carboniferous cover, Skiddaw Slates, for example, being recorded in the Crook Borehole in Co. Durham (Woolacott, 1923).

(c) CAUTLEY AND DENT INLIERS (Column N5)

The Cautley area is the type area for the Ashgill Series (Marr 1913, 1916; Ingham 1966, 1970; Ingham & Wright 1970), and has recently been remapped by Ingham (1966) who has revised the faunal sequence in detail. A small outcrop of andesite, ascribed to the Borrowdale Volcanic Group, is succeeded by the Cautley Mudstones, a succession of calcareous mudstones with occasional sandy beds and with the acidic Cautley Volcanic Formation near the top. The Cautley Mudstones range in age from the *Onnia superba* Zone of the Onnian Stage to Rawtheyan.

A stratigraphic break separates the Cautley Mudstones from overlying beds which have been correlated with the Hirnantian Stage of the Bala district (Ingham & Wright 1970; Ingham 1970). The Ashgill Shales, which are thicker here than in the Lake District, consist of distinctive grey mudstones with a sandy and sometimes conglomeratic horizon near the top; they contain a *Hirnantia* fauna. At the base is the thin Cystoid Limestone which was grouped with the Ashgill Shales (as Zone 8) by Ingham (1966) because the two units contain closely allied species of *Dalmanitina*. However, other faunal evidence from the Cystoid Limestone suggests that this unit (and therefore the '*Phacops mucronatus* Beds' of the Lake District) may differ little in age from the upper Drummuck Beds of Girvan which have been regarded as late Rawtheyan in age (Ingham 1966, p. 495; Ingham & Wright 1970, p. 237). The latter are younger than Ingham's Zone 7 and may partly equate with the Cystoid Limestone which could itself therefore be of latest Rawtheyan age. Such an assessment depends on comparison with the type Hirnantian Foel-y-Ddinas Mudstones of Bala. According to Temple (1952, p. 23) these beds probably contain *Dalmanitina olini* Temple, a species found also in the

Cystoid Limestone, although Whittington (1962-68, p. 98) was unable to give an unequivocal identification to his Bala specimens. Recent collecting at, or from near the mapped base of the Foel-y-Ddinas Mudstones on Foel-y-Ddinas (Bassett *et al.* 1966) has yielded, besides *Dalmanitina* sp., a well preserved brachiopod fauna made up of typical *Hirnantia* faunal elements—not like the brachiopod fauna of the Cystoid Limestone which has been compared with earlier Rawtheyan faunas (Ingham & Wright 1970, p. 239). The Cystoid Limestone and its north of England equivalents have therefore been placed at the top of the Rawtheyan Stage on the chart and not in the Hirnantian as hitherto.

(D) CRAVEN INLIERS (Column N 6)

In the Craven Inliers of Ribblesdale, Crummackdale and Ingleton, members of the Coniston Limestone Group rest unconformably on the Ingleton Group [Ingletonian Series]. This sequence of unfossiliferous feldspathic grits and mudstones is generally regarded as being Pre-Cambrian in age (King 1932; King & Wilcockson 1934; Leedal & Walker 1950), although there is some radiometric data which suggests that the group may be as young as early Ordovician (O'Nions *et al.* 1972).

The faunal lists of King & Wilcockson (1934, pp. 10–11) for the Douk Ghyll Mudstones and Crag Hill Limestone suggest (Ingham 1966, pp. 491–2) a Cautleyan age for these beds. Recent examination of King & Wilcockson's collections by the authors has confirmed this suggestion and indicates a mid Cautleyan age. The Horton Neptunean Dyke, an isolated outcrop of fossiliferous limestone in a fissure in rocks of the Ingleton Group, has been discussed in detail by King (1932) who suggested a correlation with the Keisley and Chair of Kildare Limestones. As commented above, Wright (1968) and Whittington (1968) regard these limestones as fairly high Ashgill which would imply a similar age for the neptunean dyke; but Ingham (1966, p. 492) suggests that there need be little difference in age between the neptunean dyke and the nearby Douk Ghyll and Crag Hill beds. Many of the Horton brachiopod genera occur in the Cautleyan Portrane Limestone of Co. Dublin but some, e.g. *Brachymimulus,* true *Cliftonia,* and *Streptis* in abundance, do not. These forms are, however, characteristic of the Kildare Limestone, with the *Cliftonia* present also in the *Hirnantia* fauna at Kildare. Undoubtedly, the Horton limestone contains another calcareous facies fauna and detailed palaeontological work is necessary before precise correlation can be established. It is, therefore, only provisionally associated here with the other Cautleyan beds.

The calcareous mudstones of Austwick Beck and elsewhere contain a Rawtheyan (Zone 6) fauna (Ingham 1966), which is associated with an unusually early '*Meristina* cf. *crassa*' (King & Wilcockson 1934, p. 12). The succeeding ashes are readily correlated with the Cautley Volcanic Formation, while the overlying unfossiliferous mudstones may well belong to Zone 7 of the Rawtheyan Stage (Ingham, *op. cit.*). Rawtheyan (Zone 6) mudstones are also present in the small Jenkin Beck Inlier, near Ingleton (Ingham 1968, p. 309).

The Cystoid Limestone and much of the Ashgill Shales are absent, Rawtheyan beds being overlain unconformably by the Wharfe Conglomerate which is correlated by Turner (1961, p. 36) and Ingham (1966) with the conglomerate within the Ashgill Shales at Cautley. Above the Wharfe Conglomerate, the thin representative of the Ashgill Shales contains all the six brachiopod genera regarded by Wright (1968, Table 3) as constituting the principal elements of the *Hirnantia* fauna.

(E) ISLE OF MAN
(not included in the chart)

The thick and structurally complex Manx Slate Group of the Isle of Man has been most recently described by Simpson (1963) who recognised eleven formations having a total thickness of 7600 m and including slates, flags, quartzites, slump breccias and some volcanic rocks. Neither base nor top to the group is known and on the basis of specimens of *Dictyonema* from the upper part of the group (*op. cit.*, pp. 370, 373, 399) Simpson believed the Manx Slates to be of Cambrian age. Subsequently, Downie & Ford (1966) have described chitinozoans, acritarchs and scolecodonts from the Lonan Flags, in the lower part of the group, which they consider indicative of an early Arenig age (late Tremadoc cannot be entirely excluded). The implication is that over 6000 m thickness of the Manx Slate Group may be of Arenig age, thus forming a direct correlative of the lower part of the Skiddaw Slate Group of the Lake District.

7. Scotland

By HARRY BLACKMORE WHITTINGTON

THE early Ordovician limestones and dolomites of north-western Scotland are remarkably similar, both in rock types and faunas, to strata of the same age in Spitsbergen, eastern Greenland and western Newfoundland. In complete contrast are the Ordovician rocks of the Girvan area and the northern belt of the Southern Uplands, which include volcanic rocks, shallow water calcareous and sandy deposits, and thick greywackes. South of this belt inliers in folded Silurian rocks (the central belt of the Southern Uplands) reveal a thin black shale and mudstone sequence of Upper Ordovician rocks, resting on cherts and volcanics. These southern Scottish Ordovician rocks have their counterparts in northern Ireland, and the shallow water faunas are like those of the central and southern Appalachians of eastern North America.

Detailed mapping and faunal studies in this century have been predominantly in the Girvan area and the western part of the Southern Uplands. The majority of these studies, and older work, have been admirably summarised by Walton (*in* Craig 1965); the *Lexique Stratigraphique International* (Whittard 1960) gives the type locality, history and other details of stratigraphical divisions. The following notes amplify and modify these accounts.

(A) NORTHWEST HIGHLANDS (Column S 1)

The three highest formations of the Durness Group, here regarded as Ordovician in age, are exposed in sequence only in the Durness area in the extreme north. Thicknesses of formations are taken from Swett (1969, table 2), though these figures differ considerably from those given by Gobbett & Wilson (1960), which were adopted by Walton (*in* Craig 1965, table 5.1). Collections from these rocks in the Geological Survey Museum include Porifera, Mollusca, a brachiopod and a trilobite (personal communication, Dr. A. W. A. Rushton). Cephalopods described from this portion of the Durness Group include *Piloceras invaginatum* Salter, *P. hornei* Ulrich, Foerste & Miller (Ulrich *et al.* 1943, pp. 20, 23), *Endoceras? baculoides* (Blake), *E.? durinum* (Blake), and *Protocycloceras mendax* (Salter) (Ulrich *et al.* 1944, pp. 84, 92, 94). Ulrich *et al.* (1944, p. 23) list additional cephalopods from the Durness Group, and regard them as of middle or upper Canadian age. Gastropod opercula collected by R. A. Fortey from the Croisaphuill Formation, north shore of Loch Borralaidh, were examined by Dr. E. L. Yochelson, U.S. Geological Survey. Dr. Yochelson (personal communication, 1969) identified one of them as *Ceratopea billingsi* Yochelson, a species common to western Newfoundland and eastern Greenland, and considered the age to be high Canadian. The pygidium of *Petigurus* in the Geological Survey Museum apparently also comes from this formation (Peach *et al.* 1907, pp. 386, 629), and indicates similar age and faunal relationships; upper Canadian is correlated with the Arenig (Whittington *in* Zen *et al.* 1968, table 4–2). Higgins (1967) studied conodonts from the Durine Formation, and suggested that the highest part may be of early Llanvirn age. However, field-work by R. A. Fortey (personal communication) has suggested that the outcrops from which Higgins derived his material may be in the Croisaphuill Formation. Some part of the highest Durness Group may thus prove to be of Llanvirn age, but further investigation is needed.

On the island of Skye the Ben Suardal Limestone, highest formation of the Durness Group, is fossiliferous and considered to be equivalent to the Balnakiel and Croisaphuill formations. In the area between Skye and Durness the Ordovician formations have not been recognised.

(B) HIGHLAND BOUNDARY FAULT ZONE AND DALRADIAN ROCKS
(Column S 2)

Rocks supposedly of Ordovician age are best known in the south-western and north-eastern parts of the Highland Boundary Fault zone. In the Aberfoyle to Loch Lomond area (Jehu & Campbell 1917; Anderson 1947) there is a lower division of spilitic lavas overlain by black shales and chert; these are overlain unconformably by breccia and greywacke with a limestone. Thicknesses of these two divisions have not been given. The fossils from the older rocks were considered to be Arenig, but re-examination (*Summ. Progr. geol. Surv. G.B.* for 1962, 1963, p. 57) shows the questionable nature of the supposed graptolites, and that the brachiopods are not indicative of other than Lower Palaeozoic age. Supposed chitinozoans from black shales at Aberfoyle 'if authentic, suggest an age of Llanvirn or younger' (Downie *et al.* 1971). Farther south-west along the fault zone, on the island of Arran, black shales, chert and spilitic lavas, 300 m thick, have been thought to be of Arenig age (Anderson & Pringle 1944). However, the evidence was the presence of two inarticulate brachiopods in black shale and the similarity of the sequence to that at Aberfoyle. The shales have yielded no palynological evidence of age (Downie *et al.* 1971). The rocks above the unconformity at

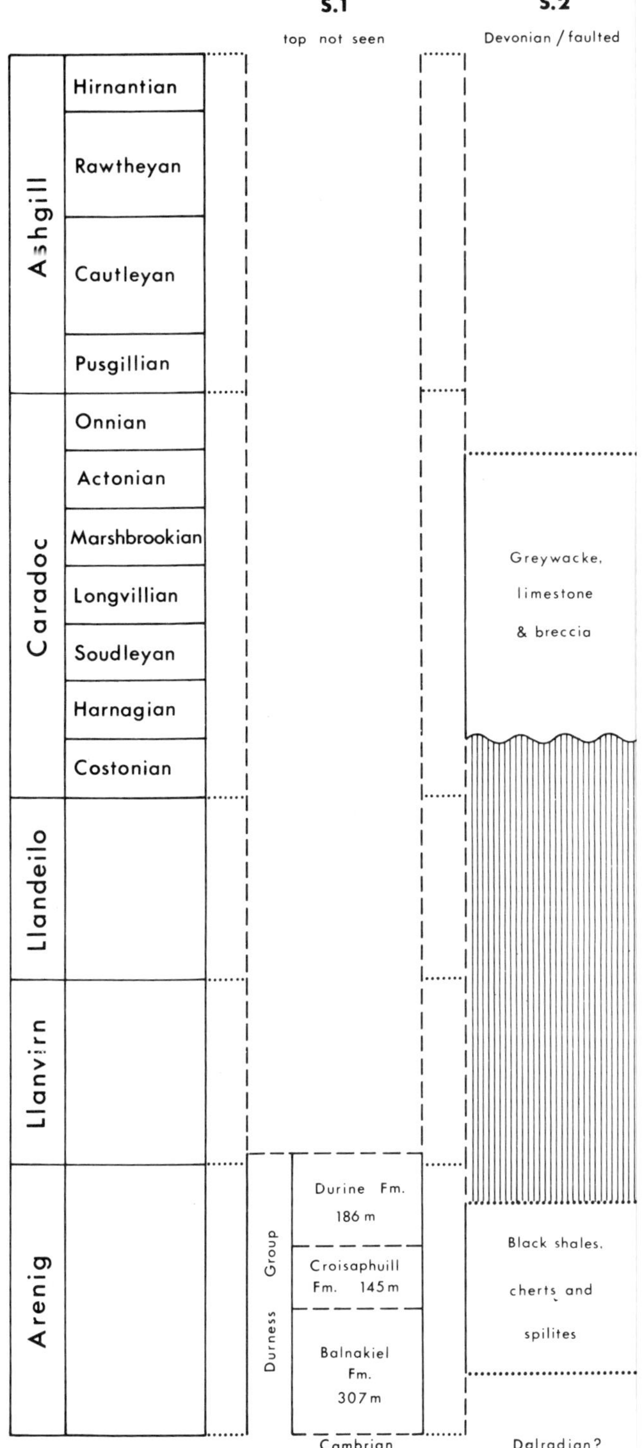

FIG. 9—Chart showing a correlation of selected Ordovician su

Dicellogr. complanatus fauna is present near the top (Toghill 1970b, p. 14). The Shalloch Formation is a flysch-type sequence, and the higher portion has yielded *Dicellogr. anceps* Nicholson (Toghill 1970b, p. 12, text fig. 4a). The Drummuck Group includes sandstones, calcareous sandstones and mudstones; the lowest shelly faunas are those of the early Cautleyan Stage. The High Mains Sandstone (Lamont 1935) has yielded the brachiopods *Hirnantia* and *Cryptothyrella*, and the trilobite *Flexicalymene*. It is exposed as a thin lens beneath unconformable Lower Llandovery rocks, and is regarded by Ingham & Wright (1970) as Hirnantian in age.

(D) NORTHERN BELT OF SOUTHERN UPLANDS (Columns S 4 and S 5)

S 4. This column is based on Walton (*in* Craig 1965, pp. 178–9, fig. 5.6), and represents the sequence immediately south of the Girvan area; a similar column has been recognised to the west, at the northern end of the Rhinns of Galloway. The Glen App Formation includes conglomerates, greywackes and siltstones, without fossils. The overlying Tappins Group (an inadequately defined name for rocks which are considered by Walton to be possibly lateral equivalents of the Dalreoch and Traboyack divisions of Williams, 1962) includes greywackes, mudstones and chert, and volcanic rocks occur at different levels within it. The Albany rocks (Williams 1962), although recognised in a small area only, are important in containing fossiliferous mudstones indicating an age equivalent to the Upper Stinchar Limestone of the Barr Group.

S 5. South-south-eastwards across the belt certain parts of the sequence are thinner, and a representative recently studied section is that at Morroch Bay (Kelling 1961, pp. 42–4). The lowest beds in this sequence have yielded conodonts (Lamont & Lindström, 1957) of Llandeilo age, and are overlain by black shales yielding graptolites of the *Nemagr. gracilis* Zone. Higher in the black shales are graptolites of Upper Glenkiln to Lower Hartfell age, and these shales give way upwards to thick greywackes of the Portpatrick Formation. The upper limits of this formation are not known, but Kelling considers that it might extend into the Silurian.

These two sections may be taken as representing the age and nature of the sediments at the western end of the northern belt of the Southern Uplands. Elsewhere in this belt there has been little recent work. That of Lamont & Lindström (1957) demonstrated that some of the cherts associated with volcanic rocks were of Arenig age; but at localities in the central and eastern part of the belt they appear to be of Llanvirn to Llandeilo age (Bergström 1971b, p. 112). Other volcanic horizons in this belt appear to be of early Caradoc age. Inter-bedded with the greywackes of this northern belt are not only thin bands of black shale with graptolites, but also conglomerates. Some of these latter are calcareous, and for example near Kilbucho, south-west of Peebles (Walton *in* Craig, 1965, p. 182) is a brachiopod fauna which is of Balclatchie age. In the valley of the Tweed southwest of Peebles is a section some 24 m thick of lavas, ashes, tuffs and limestone breccia, the Wrae Limestone (Eckford & Ritchie, 1931). Overlying these rocks are the greenish-grey Stobo Slates, 90–120 m thick. Shales immediately beneath these slates have yielded graptolites of the *clingani* Zone. From the

Wrae Limestone corals, echinoderms, brachiopods and trilobites have been obtained, believed (Pringle, 1961, p. 27) to be of early Caradoc age. However, Lindström (*in* Walton *in* Craig 1965, p. 182) studied conodonts from the limestone and considered that they suggested a pre-Caradoc age. This view is supported by Bergström (1971b, p. 112–3), whose conodont samples from the Drumelzier, Glencotho and Wrae limestones of the Tweed valley are considered by him to be upper Llanvirn in age.

(E) CENTRAL BELT OF SOUTHERN UPLANDS (Column S 6)

In the Moffat area in the centre of this belt are several sections showing the black shales, cherts and ashy mudstones of the Glenkiln Shales, and the overlying Hartfell Shales (Toghill 1970b). The latter comprises a lower division of 14 m of black, pyritic, graptolitic mudstones, and an upper division of equal thickness of barren grey mudstones with a few thin graptolite bands (*complanatus* and *anceps* Zones). Elsewhere in the central belt similar sections are revealed in the cores of anticlines.

The approximate positions of the graptolite zones recognised in this belt are shown relative to the stages of the Caradoc and Ashgill. The positions of the lower three zones follow Dean (1958), those of the upper three are after Ingham & Wright (1970).

8. Ireland

By ALWYN WILLIAMS

THE Ordovician rocks of Ireland belong to three distinctive belts reflecting the differentiation in Britain. From Galway north-eastwards to Tyrone, successions are characterized by shelly assemblages which are closely related to those found in the Ordovician sediments of eastern North America and Ayrshire. Along the Down-Longford axis, the Ordovician consists of graptolitic shales with greywackes comparable with those of the Southern Uplands. South of this axis, thick successions yielding Anglo-Welsh and Baltic shelly faunas prevail.

(A) MAYO AND GALWAY (Columns I 1 and I 2)

A number of lithostratigraphic units, exposed between Lough Mask and Killary Harbour, have been given various Formational, Group and Series names (Kilroe 1907; Carruthers & Maufe 1909; Gardiner & Reynolds 1909, 1910, 1912, 1914; Theokritoff 1951; Stanton 1960; McKerrow & Campbell 1960). The Series names (Owenmore, Murrisk, Glenummera, Partry) are inappropriate and have been discarded. Dewey (1963), in consideration of the almost uninterrupted nature of the rock successions of Central Murrisk, suggested that the sections in that district should be adopted as type for the region. His recommendation is accepted here although the scarcity of fossils within the successions makes correlation imprecise even with the more fossiliferous rocks of the Partry Mountains immediately to the east.

I 1. The Ordovician of Central Murrisk consists of slide conglomerates, turbidites, slates and tuffs (Letterbrock to Glenummera Groups inclusive), followed by grits, conglomerates and breccias with ignimbrites and slate bands (Mweelrea and Maumtrasna Groups). *Didymogr. extensus* assemblages, representing the Zones of *Didymogr. deflexus, Didymogr. nitidus* and *Isogr. gibberulus*, have been found sporadically in the Owenmore and Sheeffry Groups and their correlatives (Dewey *et al.* 1970). Equally well defined stratigraphically is the Derrylea Group which was erected by Dewey in 1963 for the upper 1500 m of the 'Sheeffry Grits' as understood by Stanton (1960). Graptolites collected by Carruthers & Maufe (1909) from slates exposed at the north end of Doolough must have come from that part of the 'Sheeffry Grits' assigned by Dewey (1963) to the Derrylea Group. The graptolites are generally assumed to be Arenigian in age although Skevington (1971b) maintains that they cannot be older than the *Didymogr. hirundo* Zone. In any event, the implication of Dewey's restriction of the stratigraphic range of the Sheeffry Group has been overlooked in the Murrisk correlation proposed by Dewey *et al.* in 1970 (Prof. D. Skevington, personal communication) where the entire Derrylea Group is assigned to the Llanvirn. In the absence of evidence supporting this assignment, the Derrylea Group has been retained within the Arenig.

This arrangement is consistent with the age of a shelly fauna recovered from slates in the lower Mweelrea Group exposed at Uggool (Williams 1972a). The assemblage is diagnostic of the *Rhysostrophia* subzone of the Whiterock Stage (Ross & Ingham 1970) which is probably equivalent to high Llanvirn in the standard succession. The time-stratigraphic range of the Mweelrea and Maumtrasna Groups is unknown, and only the fact that they are over 5500 m thick has prompted their classification as at least early Llandeilian.

I 2. In the eastern part of the region, in the vicinity of Loughs Mask and Nafooey, the Mweelrea Group succeeds a variety of older Ordovician rocks and its base must represent an unconformity. Dewey (1963, p. 330) proposes that the oldest Arenig rocks exposed here, which consist of spilites, acid tuffs, cherts and shales (with a *Didymogr. extensus* fauna) and were included by Gardiner & Reynolds (1912, p. 97) in their Mount Partry Beds, should be named the Lough Nafooey Group. The composition of pebbles indicates that conglomerates within the Owenmore Group were derived from the Lough Nafooey Group which is recognised to be the oldest Arenig succession in the area and is assumed to be separated from the succeeding Arenig deposits everywhere by an unconformity. In the Lough Mask area, the succession intervening between the base of the Mweelrea and Lough Nafooey Groups consists of conglomerates, grits, shales, breccias, limestones and ashes. They constitute part of the Mount Partry Beds and the Shangort and Tourmakeady Beds of Gardiner & Reynolds (1909, 1910) and include correlatives of the Leenane Grits (McKerrow & Campbell 1960). In recognition of the fact that the entire succession is best exposed and understood in the Glensaul area, it is proposed to include these rocks in the Glensaul Group and to restrict the stratigraphic terms established by Gardiner & Reynolds and McKerrow & Campbell to more localised mappable units. The limestones, which form a series of impersistent lenticles and breccias within the upper part of the succession, are richly fossiliferous with an orthid-finkelnburgiid-porambonitacean brachiopod fauna and a solenopleurid-bathyurid-cheirurid trilobite assemblage of American affinities (Williams 1969b). In Glensaul, these faunas

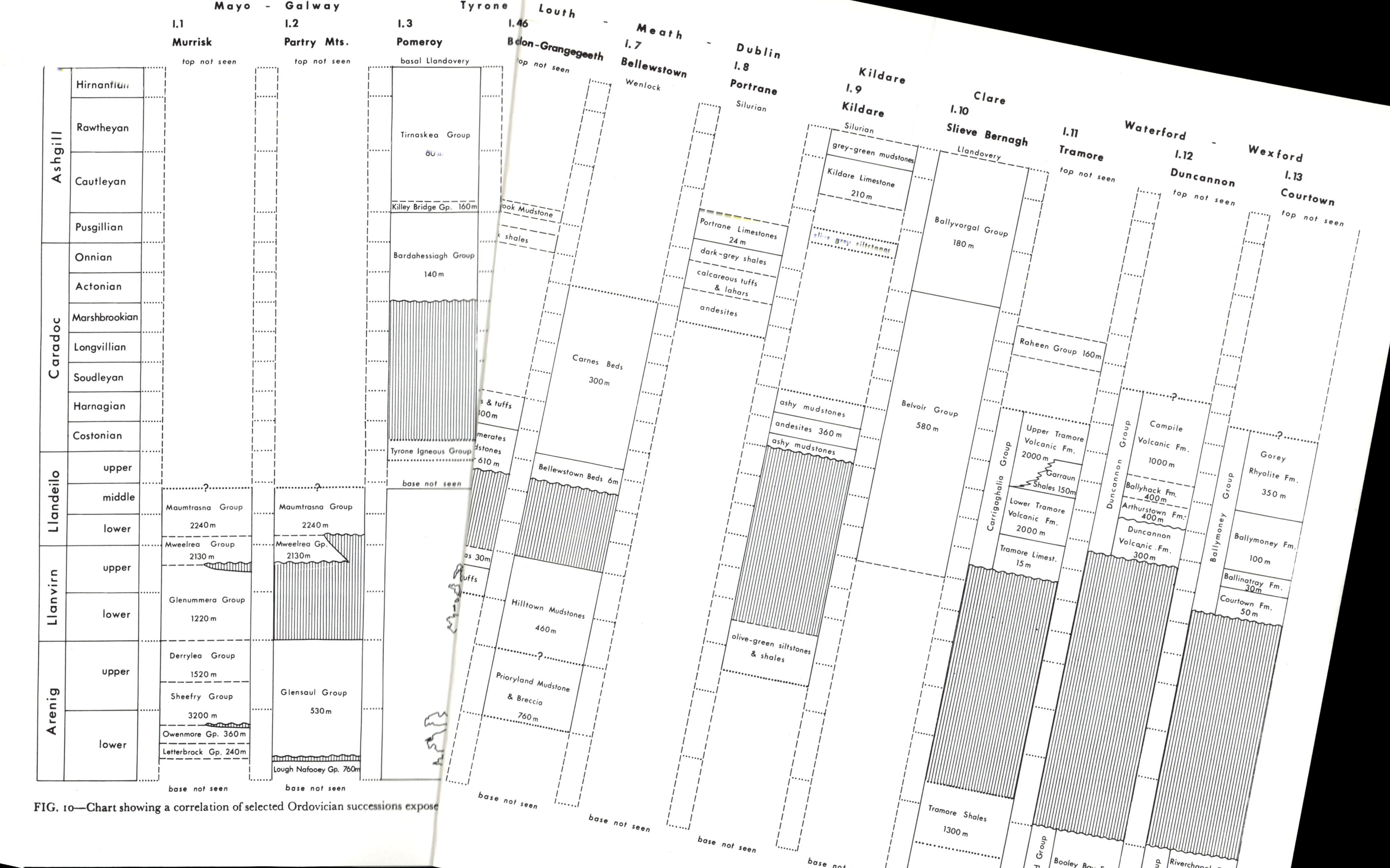

FIG. 10—Chart showing a correlation of selected Ordovician successions expose

occur between two beds of shales containing graptolites diagnostic of the *Isogr. gibberulus* and *Didymogr. hirundo* Zones (Dewey *et al.* 1970), and a late Arenigian age is also currently determinable for the brachiopod fauna. Whittington (1968, p. 56) considers the Group to be somewhat younger because he found the trilobite assemblage to be very like that characteristic of the Whiterock Stage (Llanvirn) of North America. There is, therefore, some conflict in age determination although this may ultimately prove to be a matter of trans-Atlantic nomenclature rather than any significant time-stratigraphic difference. Provisionally, the entire fauna is more acceptable as late Arenig and, since it is found in the upper part of the Glensaul Group, it is possible that the pre Mweelrea break spans most of the Llanvirn.

(B) DOWN, TYRONE AND CAVAN (Columns I 3 to I 5)

The Ordovician successions of Down, Tyrone and Cavan, like the structurally and faunally related contemporaneous rocks of the Southern Uplands of Scotland, are involved in a strong facies change across the strike.

I 3. In the Pomeroy region, the Ordovician sediments bear rich shelly assemblages with some graptolites (Fearnsides *et al.* 1907; Reed 1952; Mitchell 1969), but their actual relationship with the Tyrone Igneous Group is not seen in the field. The general stratigraphic setting of the Igneous Group (Hartley 1933), which consists of an andesitic-rhyolitic suite of agglomerates and lavas with some chert and black shales, is like that of the Arenig Ballantrae Igneous Group. However, *Dicranogr. rectus* (Hopkinson) has been collected from the shales (Hartley 1936, p. 227) and indicates contemporaneity with the volcanic rocks of Ballygrot. The Bardahessiagh Group, composed of siltstone and sandstones, probably sits unconformably on the Tyrone Igneous Group because loose blocks of fossiliferous conglomerates have been found in the vicinity of the inferred junction. The Bardahessiagh shelly fauna is closely comparable with that found in the Upper Ardwell-Craighead successions of the Girvan area. The Killey Bridge micaceous siltstones and mudstones yield shelly fossils which are characteristic of the Lower Cautleyan. They are succeeded by the siltstones and mudstones of the Tirnaskea Group which have yielded *Dalmanitina* indicating the presence of at least Hirnantian horizons.

I 4. In North Down, the oldest rocks exposed in the structural equivalent of the Northern Belt of the Southern Uplands (Sharpe 1970), consist of unfossiliferous shales succeeded by pillow lavas with minor cherts and shales which are in turn followed by tuffaceous agglomerates and shales, cherts and greywackes with a graptolitic fauna reminiscent of both the *Nemagr. gracilis* and *Climacogr. peltifer* Zones.

I 5. The Coalpit Bay succession (Swanston 1876-7; Lapworth 1876-7) is comparable with that found in the Central Belt of the Southern Uplands (Hartley & Harper 1937). It is badly faulted but lithology and graptolitic faunas suggest that a complete Glenkiln-Hartfell succession was originally present. Clark (1902, p. 498) recorded *Dicellogr. anceps* from the pale grey mudstones, and officers of the Institute of Geological Sciences have identified the *Climacogr. peltifer* and *Climacogr. wilsoni* Zones from the underlying black shales as well as a *Dicranogr. clingani* or *Pleurogr.*

linearis fauna from shales exposed nearby at Carnalea (Dr. A. W. A. Rushton, personal communication).

The Lough Acanon Pelitic Formation, which includes dark graptolitic shales, cherts and basic volcanic rocks, is only sporadically exposed in Co. Cavan, but is believed by Phillips & Skevington (1968) to embrace a complete succession of rocks between the *Climacogr. wilsoni* and *Akidogr. acuminatus* Zones.

(c) LOUTH, MEATH AND DUBLIN (Columns I 6 to I 8)

Although the successions in the Ordovician inliers south of Dundalk range throughout the System (Harper 1952, Harper & Rast 1964, Brenchley, Harper, Romano & Skevington 1967), the sporadic distribution of fossils and poorness of exposures preclude the drawing of accurate time-stratigraphic boundaries.

I 6, I 7. The Prioryland Mudstones and Breccias are unfossiliferous and are assigned to the Arenig only because they are older than mudstones, bedded ashes, ignimbrites and mud flows (the Hilltown Mudstones) which have yielded graptolites of the *Didymogr. bifidus* Zone. The calcareous shales and limestones comprising the Bellewstown Beds have been assigned to the Llanvirn by Harper & Rast (1964, p. 18), but newly made collections of the shelly fauna found in the Formation suggest correlation with the Tramore Limestone (W. I. Mitchell, personal communication). At Bellewstown, the Carnes Beds, which consist of shales, mudstones and calcareous ash, yielded a graptolitic fauna indicative of the *Diplogr. multidens* Zone below and a shelly fauna typical of the Actonian above. Correlatives in the Collon-Grangegeeth area include the tuffs and shales of the Grangegeeth Volcanic Group which contain a shelly fauna with Baltic affinities. Harper (1952, p. 86) originally emphasised the Balclatchie likeness of the fauna but more recently (with Rast 1964, p. 18) has assigned it to the Costonian. New collections reveal the presence of many species known in the Tramore Limestone (W. I. Mitchell, personal communication) and confirm that the fauna is older than was first thought. The stratigraphic range of these sandstones, tuffs and shales is therefore greatly extended because the upper shales have yielded 'Balclatchie' trilobites and graptolites consistent with the Zone of *Climacogr. peltifer* (Brenchley, Harper, Romano & Skevington 1967, p. 300). Younger Ordovician sediments in the area include black shales with a *Pleurogr. linearis* fauna and the Oriel Brook Mudstones with a trilobite assemblage like that of the Upper Whitehouse Beds. Their relationships to one another and to older successions are unknown.

I 8. The succession at Portrane (Gardiner & Reynolds 1897) is noteworthy because it includes the Portrane Limestone with its rich silicified brachiopod, coral, bryozoan and ostracode faunas (Wright 1963, 1964, Kaljo & Klaamann 1965, Ross 1966). A. D. Wright who is currently completing the stratigraphy and correlation of the succession concludes that the Portrane Limestone is equivalent to high Cautleyan. The older rocks are much less accurately determinable. The underlying dark grey shales contain trilobites like *Tretaspis portrainensis* (Reed) (Lamont 1941, p. 453) which is unlikely to be earlier than Onnian, while the calcareous tuffs with lahars have yielded to Wright a restricted Upper Ordovician assemblage of *Rostricellula*, *Plaesiomys* and *Eoplectodonta*. Consequently the dating of these formations on the chart is provisional.

(D) KILDARE (Column I 9)

I 9. The Ordovician succession at the Chair of Kildare, which has been described by Reynolds & Gardiner (1896), is best understood through the researches of A. D. Wright (1967, 1968, 1970 and MSS). Olive-green mudstones with siltstone bands are the oldest exposed rocks and probably the source of *Didymogr. bifidus* recorded by Elles & Wood (1901-18, p. 43). Lava flows, principally of andesitic composition are underlain by a variable suite of ashy sediments containing a Soudleyan shelly fauna, and are capped by ashy olive-grey siltstones with a Longvillian orthoid-strophomenoid brachiopod fauna (Wright 1970). The reefal Kildare Limestone flanked by olive-green and red mudstones are richly fossiliferous with brachiopod-trilobite assemblages indicative of a late Ashgillian age. A *Hirnantia* fauna has been identified by Wright (1968) in the topmost member of the Limestone, with the underlying reefs probably ranging down to the middle Rawtheyan. The Limestone is underlain by olive mudstones and shales of unknown range and apparently terminated by grey-green mudstones which are reported to have yielded a trinucleid (Reynolds & Gardiner 1896, p. 594). As Wright points out the trinucleid, if correctly located, is almost certainly a *Tretaspis*, and its presence implies that either the *Hirnantia* fauna of Kildare is older than that of Britain or the *Tretaspis*-bearing mudstone is a facies variant of the topmost *Hirnantia* beds in Britain.

(E) CLARE (Column I 10)

I 10. A reasonably complete succession of Upper Ordovician rocks is found in Slieve Bernagh (Weir 1962) although only two datable fossiliferous horizons are known. In the Belvoir Group, 80 m of black shales bearing a *Nemagr. gracilis* fauna intervene between 425 m of unfossiliferous cherts, grits and mudstones below and unfossiliferous horizons above. The siltstones are succeeded conformably by the mudstones of the Ballyvorgal Group which, a metre or so above the base, have yielded a Bohemian trilobite fauna close to that characteristic of the Upper Whitehouse Group of Girvan (Lower Cautleyan).

(F) WEXFORD AND WATERFORD (Columns I 11 to I 13)

The greatest area of Ordovician outcrops occurs in south-east Ireland, but successions are poorly understood and, except for a few like those found in the Tramore, Duncannon and Courtown districts, cannot be arranged with any confidence. Arenig rocks with graptolitic faunas (Harper 1948, p. 62; Brenchley, Harper & Skevington 1967, p. 389) are known to occur in Enniscorthy and Arklow where a succession of Ordovician slates, sandstones and volcanics is nearly 8,000 m thick according to Tremlett (1959, p. 19), while a shelly fauna from the Tagoat Beds has been described by Brenchley *et al.* (1967, p. 387). Caradocian graptolitic and shelly faunas also occur in shales and sandstones in a number of localities including Slieve Roe (Co. Wicklow), Enniscorthy and Carrickdaggan; and it is likely that a wide range of Ordovician rocks is represented by the belts of siltstones, grits, slates and volcanics flanking the Leinster granite.

I 11. The succession at Tramore has been described by Reed (1899), C. J. Murphy (*in* Brindley & Gill 1958, p. 256) and especially by Mrs. H. Carlisle

(1971). The Tramore Shale Formation with minor siltstones and mudstones is unfossiliferous but has been assigned to the Arenig in the belief that it is older than a substantial Llanvirn-Llandeilo break which seems to be prevalent in south-east Ireland. The younger Carrigaghalia Group begins with the Tramore Limestone consisting of dark blue argillaceous and arenaceous limestones and containing a rich shelly fauna of Baltic affinities. The basal shale member of the conformable Lower Tramore Volcanic Formation contains *Nemagr. gracilis.* *Diplogr. multidens* has also been recorded from other shales within the Formation which consists mainly of rhyolites and andesites. No fossils have yet been recovered from impersistent shales and siltstones (Garraun Shales) separating the Upper Tramore Volcanic Formation of rhyolite from the Lower, but correlation with the Duncannon sections suggests that the Garraun Shales may be about Longvillian in age. The Raheen Group also consists of rhyolitic flows, agglomerates and mudstones, but the succession is isolated from older rocks and, judging from its restricted fauna which includes *Onniella* and *Tretaspis ceryx* (Lamont), it is late Caradocian to early Ashgillian in age.

I 12. The succession in the vicinity of Duncannon (Gardiner 1967; Gardiner & Brenchley 1970) differs from that at Tramore in a number of respects. The siltstones and shales composing the Booley Bay Formation are locally unfossiliferous but are demonstrably part of the Ribband Group which have yielded graptolites indicative of the Arenig at Breanoge Head and Kiltrea (Brenchley, Harper & Skevington 1967, p. 369). The junction between the Booley Bay Formation and the succeeding Duncannon Group is not exposed although it must be a profound unconformity. The Duncannon Group is dominated by two volcanic suites: the andesites, dacites and derived rocks forming the older Duncannon Volcanic Formation and the rhyolites of the younger Campile Volcanic Formation. However, shales within the top part of the Duncannon Volcanic Formation have yielded graptolites (including *Orthogr.* cf. *calcaratus* Lapworth) diagnostic of the *Climacogr. peltifer/Diplogr. multidens* Zones, while another graptolite assemblage recovered from the top of the siltstones and shales constituting the Arthurstown and Ballyhack Formations includes *Diplogr. compactus* (Elles & Wood) which is suggestive of the *Dicranogr. clingani* Zone. If the Campile and Upper Tramore Volcanic Formations are correlatable, the Arthurstown and Ballyhack Formations must have been at least partly contemporaneous with the Garraun Shales which is therefore about Soudleyan-Longvillian in age.

I 13. The Ordovician rocks exposed in the vicinity of Courtown suggest that the Caradocian sequence found in the Tramore area is more characteristic of south-east Ireland than is generally believed (Mitchell *et al.* 1972). The older rocks, allocated by Crimes & Crossley (1968) and Brenchley & Treagus (1970) to the Ribband Group, have been divided into a number of Formations. However, only the three youngest are fossiliferous having yielded graptolites diagnostic of the Lower Arenig and these alone are here included in the Ordovician. The Ballymoney Group of Brenchley & Treagus succeeds the Ribband Group with a strong unconformity because a rich fauna from the calcareous shales and silty limestones of the Courtown Formation is coeval with that of the Tramore Limestone. As at Tramore, these carbonate beds are followed by shales yielding *Nemagr.*

gracilis (Ballinatray Formation), while shelly assemblages from the ashes and tuffs of the Ballymoney Formation are likely to be of Harnagian-Soudleyan age (Mitchell *et al.* 1972). The time-stratigraphic range of the Gorey Rhyolite Formation is unknown but cannot exceed more than a Stage or so of the Caradocian.

9. References

The numbers in square brackets at the end of many of the references refer to areas mapped in Wales and the Borderland and shown in Fig. 4.

ADAMS, T. D. 1963. The geology of the Dinas Cwm-Rheidol hydroelectric tunnel. *Geol. Mag.* **100**, 371–8.

AMERICAN COMMISSION ON STRATIGRAPHIC NOMENCLATURE. 1961. Code of Stratigraphic Nomenclature. *Bull. Am. Ass. Petrol. Geol.* **45**, 645–65.

ANDERSON, J. G. C. 1947. The geology of the Highland Border: Stonehaven to Arran. *Trans. R. Soc. Edinb.* **61**, 479–515.

—— 1963. The geology of the Rheidol hydroelectric project—Cardiganshire. *Proc. S. Wales Inst. Engrs* **78**, 35–47.

—— & PRINGLE, J. 1944. The Arenig rocks of Arran, and their relationships to the Dalradian Series. *Geol. Mag.* **81**, 81–7.

ANDREW, G. 1925. The Llandovery rocks of Garth (Breconshire). *Q. Jl geol. Soc. Lond.* **81**, 389–406. [92].

ANKETELL, J. M. 1963. *The geology of the Llangranog district, southwest Cardiganshire.* Thesis (Ph.D.) Belfast. [85].

APOLLONOV, M. K., BANDALETOV, S. M., NIKITIN, I. F. & TZAY, D. T. 1968. Kazakhstan Ordovician and Silurian deposits and their correlation with European columns. *Acad. Nauk. U.S.S.R.*, 23rd *Int. geol. Congr., Doklady Soviet. geol.* **9**, 111–23. (In Russian, English summary.)

BAKER, J. W. 1966. The Ordovician and other post-Rosslare Series rocks in south-east Co. Wexford. *Geol. J.* **5**, 1–6.

BANCROFT, B. B. 1928. On the unconformity at the base of the Ashgillian in the Bala district. *Geol. Mag.* **65**, 484–93. [48].

—— 1933. *Correlation tables of the Stages Costonian-Onnian in England and Wales.* Blakeney, Glos. (privately printed), 1–4.

—— 1945. The brachiopod zonal indices of the Stages Costonian to Onnian in Britain. *J. Paleont.* **19**, 181–252.

BASSETT, D. A. 1961. *Bibliography and index of geology and allied sciences for Wales and the Welsh Borders 1897–1958.* National Museum of Wales, Cardiff, 1–376.

—— 1963. *Bibliography and index of geology and allied sciences for Wales and the Welsh Borders 1536–1896.* National Museum of Wales, Cardiff, i–x, 1–246.

—— 1967. *A source-book of geological, geomorphological and soil maps for Wales and the Welsh Borders (1800–1966).* National Museum of Wales, Cardiff, i–x, 1–239.

—— 1969. Some of the major structures of early Palaeozoic age in Wales and the Welsh Borderland: an historical essay. *In* Wood, A. (Ed.) *The Pre-Cambrian and Lower Palaeozoic rocks of Wales.* University of Wales Press, Cardiff, 67–116.

——, WHITTINGTON, H. B. & WILLIAMS, A. 1966. The stratigraphy of the Bala district, Merionethshire. *Q. Jl geol. Soc. Lond.* **122**, 219–71. [50].

BATES, D. E. B. 1968. The Lower Palaeozoic brachiopod and trilobite faunas of Anglesey. *Bull. Br. Mus. nat. Hist.* (Geol.) **16**, 127–99.

—— 1969a. Some aspects of the Arenig faunas of Wales. *In* Wood, A. (Ed.) *The Pre-Cambrian and Lower Palaeozoic rocks of Wales.* University of Wales Press, Cardiff, 155–9.

—— 1969b. Some early Arenig brachiopods and trilobites from Wales. *Bull. Br. Mus. nat. Hist.* (Geol.) **18**, 1–28.

—— 1972. The stratigraphy of the Ordovician rocks of Anglesey. *Geol. J.* **8**, 19–58. [2].

BEAVON, R. V. 1963. The succession and structure east of the Glaslyn river, North Wales. With appendix: The Caradocian faunas below the Lower Rhyolitic Tuff and above the Llyn Llagi Tuff, by A. Williams and J. C. Harper. *Q. Jl geol. Soc. Lond.* **119,** 479–512. [18].

BERGSTRÖM, S. M. 1964. Remarks on some Ordovician faunas from Wales. *Acta Univ. lund.* Sec. II. No. 3, 1–67.

—— 1971a. Correlation of the North Atlantic Middle and Upper Ordovician conodont zonation with the graptolite succession. *Mém. Bur. Rech. géol. minièr.* **73,** 177–87.

—— 1971b. Conodont biostratigraphy of the Middle and Upper Ordovician of Europe and eastern North America. *Mem. geol. Soc. Am.* **127,** 83–157.

BERRY, W. B. N. 1960. Correlation of Ordovician graptolite-bearing sequences. 21st *Int. geol. Congr.*, Copenhagen, **7,** 97–108.

—— 1967. Comments on correlation of the North American and British Lower Ordovician. *Bull. geol. Soc. Am.* **78,** 419–28.

—— 1968. British and North American Lower Ordovician correlation: reply. *Bull. geol. Soc. Am.* **79,** 1265–72.

—— & BOUCOT, A. J. 1970. Correlation of the North American Silurian rocks. *Spec. Pap. geol. Soc. Am.* **102,** 1–289.

BLACK, W. W., BULMAN, O. M. B., HEY, R. W. & HUGHES, C. P. 1972. Ordovician stratigraphy of Abereiddy Bay, Pembrokeshire. *Geol. Mag.* **108** (for 1971), 546–8.

BLACKIE, R. C. 1928. The geology of the country between Llanelidan and Bryneglwys. *Q. Jl geol. Soc. Lond.* **83** (for 1927), 711–36. [51].

BRENCHLEY, P. J. 1964. Ordovician ignimbrites in the Berwyn Hills, North Wales. *Geol. J.* **4,** 43–54.

—— 1966. *The Caradoc rocks of the north and west Berwyns, North Wales.* Thesis (Ph.D.) Liverpool. [56].

—— 1969. The relationship between Caradocian volcanicity and sedimentation in North Wales. *In* Wood, A. (Ed.) *The Pre-Cambrian and Lower Palaeozoic rocks of Wales.* University of Wales Press, Cardiff, 181–202.

——, HARPER, J. C., ROMANO, M. & SKEVINGTON, D. 1967. New Ordovician faunas from Grangegeeth, Co. Meath. *Proc. R. Ir. Acad.* Sec. B. **65,** 297–304.

——, —— & SKEVINGTON, D. 1967. Lower Ordovician shelly and graptolitic faunas from south-eastern Ireland. *Proc. R. Ir. Acad.* Sec. B. **65,** 385–90.

—— & TREAGUS, J. E. 1970. The stratigraphy and structure of the Ordovician rocks between Courtown and Kilmichael Point, Co. Wexford. *Proc. R. Ir. Acad.* Sec. B. **69,** 83–102.

BRINDLEY, J. C. & GILL, W. D. 1958. Summer field meeting in southern Ireland, 1957. *Proc. Geol. Ass.* **69,** 244–61.

BROMLEY, A. V. 1965. Intrusive quartz latites in the Blaenau Ffestiniog area, Merioneth. *Geol. J.* **4,** 247–56. [17].

—— 1969. Acid plutonic igneous activity in the Ordovician of North Wales. *In* Wood, A. (Ed.) *The Pre-Cambrian and Lower Palaeozoic rocks of Wales.* University of Wales Press, Cardiff, 387–408.

BULMAN, O. M. B. 1963. On *Glyptograptus dentatus* (Brongniart) and some allied species. *Palaeontology* **6,** 665–89.

—— 1970. *Treatise on Invertebrate Palaeontology.* Part V, *Graptolithina with sections on Enteropneusta and Pterobranchia.* 2nd Ed. (Revised & Enlarged). The Geological Society of America, Inc., Boulder, Colorado & The University of Kansas, Lawrence, Kansas, i–xxxii, V1–V163.

—— 1971. Graptolite faunal distribution. *In* Middlemiss, F. A., Rawson, P. F. & Newall, G. (Eds.) Faunal provinces in space and time. *Geol. J.* Special Issue No. 4, 47–60.

BURGESS, I. C. 1965. District reports. Brough under Stainmore (31) Sheet. *Summ. Progr. geol. Surv. Lond.* (for 1964), 57–8.

CANTRILL, T. C., DIXON, E. E. L., THOMAS, H. H. & JONES, O. T. 1916. The geology around the South Wales Coalfield. Part XII. The country around Milford. (Sheet 227). *Mem. geol. Surv. U.K.,* i–vii, 1–185. [109].

—— & THOMAS, H. H. 1906. On the igneous and associated sedimentary rocks of Llangynog (Caermarthenshire). *Q. Jl geol. Soc. Lond.* **62,** 223–52. [97].

CARLISLE, H. 1971. *The geology of the Tramore area*. Thesis (B.Sc.) Belfast.

CARRUTHERS, R. G. & MAUFE, H. B. 1909. The Lower Palaeozoic rocks around Killary Harbour. *Ir. Nat.* **18**, 7–11.

CATTERMOLE, P. & JONES, A. 1970. The geology of the area around Mynydd Mawr, Nantlle, Caernarvonshire. *Geol. J.* **7**, 111–28. [22].

CAVE, R. 1955. *The stratigraphy of the Welshpool area (Montgomeryshire)*. Thesis (Ph.D.) Cambridge. [64].

—— 1957. *Salterolithus caractaci* (Murchison) from Caradoc strata near Welshpool, Montgomeryshire. *Geol. Mag.* **94**, 281–90.

—— 1960. A new species of *Tretaspis* from South Wales. *Geol. Mag.* **97**, 334–7.

—— 1965. The Nod Glas sediments of Caradoc age in North Wales. *Geol. J.* **4**, 279–98. [63].

CHALLINOR, J. 1951. Geological research in Cardiganshire 1842–1942. *Ceredigion* **1**, 145–76.

CLARK, R. 1902. Notes on the fossils of the Silurian area of north-east Ireland. *Geol. Mag.* **14**, 497–500.

COCKS, L. R. M. 1968. Some strophomenacean brachiopods from the British Lower Silurian. *Bull. Br. Mus. nat. Hist.* (Geol.) **15**, 285–324.

——, HOLLAND, C. H., RICKARDS, R. B. & STRACHAN, I. 1971. A correlation of Silurian rocks in the British Isles. *Jl geol. Soc. Lond.* **127**, 103–36.

——, TOGHILL, P. & ZIEGLER, A. M. 1970. Stage names within the Llandovery Series. *Geol. Mag.* **107**, 79–87.

COOPER, B. N. & COOPER, G. A. 1946. Lower Middle Ordovician stratigraphy of the Shenandoah valley, Virginia. *Bull. geol. Soc. Am.* **57**, 35–113.

COOPER, G. A. 1956. Chazyan and related brachiopods. *Smithson misc. Collns.* **127**, Pt. 1, 1–1024, Pt. 2, 1025–245.

COWIE, J. W., RUSHTON, A. W. A. & STUBBLEFIELD, C. J. 1972. A correlation of Cambrian rocks in the British Isles. *Geol. Soc. Lond. Spec. Rep.* No. 2, 1–42.

COX, A. H. 1916. The geology of the district between Abereiddy and Abercastle (Pembrokeshire). *Q. Jl geol. Soc. Lond.* **71** (for 1915), 273–342. [102].

—— 1925. The geology of the Cader Idris range (Merioneth). *Q. Jl geol. Soc. Lond.* **81**, 539–94.

——, GREEN, J. F. N., JONES, O. T. & PRINGLE, J. 1930. The geology of the St. David's district, Pembrokeshire. *Proc. Geol. Ass.* **41**, 241–73.

—— & WELLS, A. K. 1921. The Lower Palaeozoic rocks of the Arthog-Dolgelley district (Merionethshire). *Q. Jl geol. Soc. Lond.* **76** (for 1920), 254–324. [39].

CRAIG, G. Y. (Ed.) 1965. *The geology of Scotland*. Oliver & Boyd, Edinburgh and London, i–xvi, 1–556.

CRIMES, T. P. 1970. A facies analysis of the Arenig of western Lleyn, North Wales. *Proc. Geol. Ass.* **81**, 221–39.

—— & CROSSLEY, J. D. 1968. The stratigraphy, sedimentology, ichnology and structure of the Lower Palaeozoic rocks of part of north-eastern Co. Wexford. *Proc. R. Ir. Acad. Sec. B.* **67**, 185–215.

CROSFIELD, M. C. & SKEAT, E. G. 1896. On the geology of the neighbourhood of Carmarthen. *Q. Jl geol. Soc. Lond.* **52**, 523–41. [96].

DAKYNS, J. R. 1877. On a base to the Carboniferous rocks in Teesdale. *Proc. Yorks. geol. Soc.* **6**, 239–42.

DAVIES, D. A. B. 1936. The Ordovician rocks of the Trefriw district (North Wales). *Q. Jl geol. Soc. Lond.* **92**, 62–90. [33].

DAVIES, K. A. 1926. The geology of the country between Drygarn and Abergwesyn (Breconshire). *Q. Jl geol. Soc. Lond.* **82**, 436–64. [81].

—— 1928. Contributions to the geology of Central Wales. I.—Notes on the geology of the southern portion of Central Wales. II.—The geology of the country between Rhayader (Radnorshire) and Abergwesyn (Breconshire). *Proc. Geol. Ass.* **39**, 157–68. [80].

—— 1933. The geology of the country between Abergwesyn (Breconshire) and Pumpsaint (Carmarthenshire). *Q. Jl geol. Soc. Lond.* **89**, 172–201. [82].

—— & PLATT, J. I. 1933. The conglomerates and grits of the Bala and Valentian rocks of the district between Rhayader (Radnorshire) and Llansawel (Carmarthenshire). *Q. Jl geol. Soc. Lond.* **89,** 202–20.

DAVIES, R. A. 1969. *Geological succession and structure of Cambrian and Ordovician rocks in the northeastern Carneddau.* Thesis (Ph.D.) Wales. [26].

DAVIES, R. G. 1959. The Cader Idris granophyre and its associated rocks. *Q. Jl geol. Soc. Lond.* **115,** 189–216. [41].

DEAN, W. T. 1958. The faunal succession in the Caradoc Series of south Shropshire. *Bull. Br. Mus. nat. Hist.* (Geol.) **3,** 191–231.

—— 1959. The stratigraphy of the Caradoc Series in the Cross Fell Inlier. *Proc. Yorks. geol. Soc.* **32,** 185–228.

—— 1960a. The Ordovician trilobite faunas of south Shropshire, I. *Bull. Br. Mus. nat. Hist.* (Geol.) **4,** 71–143.

—— 1960b. The use of shelly faunas in a comparison of the Caradoc Series in England, Wales and parts of Scandinavia. 21st *Int. geol. Congr.*, Copenhagen, **7,** 82–7.

—— 1961. The Ordovician trilobite faunas of south Shropshire, II. *Bull. Br. Mus. nat. Hist.* (Geol.) **5,** 311–58.

—— 1962. The trilobites of the Caradoc Series in the Cross Fell Inlier of northern England. *Bull. Br. Mus. nat. Hist.* (Geol.) **7,** 65–134.

—— 1963a. The Ordovician trilobite faunas of south Shropshire, IV. *Bull. Br. Mus. nat. Hist.* (Geol.) **9,** 1–18.

—— 1963b. The Stile End Beds and Drygill Shales in the east and north of the English Lake District. *Bull. Br. Mus. nat. Hist.* (Geol.) **9,** 47–65.

—— 1964. The geology of the Ordovician and adjacent strata in the southern Caradoc district of Shropshire. *Bull. Br. Mus. nat. Hist.* (Geol.) **9,** 257–96.

—— 1965. A shelly fauna from the Snowdon Volcanic Series at Twll Ddu, Caernarvonshire. *Geol. J.* **4,** 301–14.

—— 1967. Relationships of the Shelve trilobite faunas. *In* Whittard, W. F. The Ordovician trilobites of the Shelve Inlier, west Shropshire. Pt. IX. *Palaeontogr. Soc.* (*Monogr.*), 308–20.

—— & DINELEY, D. L. 1961. The Ordovician and associated Pre-Cambrian rocks of the Pontesford district, Shropshire. *Geol. Mag.* **98,** 367–76.

DEWEY, J. F. 1963. The Lower Palaeozoic stratigraphy of central Murrisk, Co. Mayo, Ireland, and the evolution of the south Mayo trough. *Q. Jl geol. Soc. Lond.* **119,** 313–44.

——, RICKARDS, R. B. & SKEVINGTON, D. 1970. New light on the age of Dalradian deformation and metamorphism in western Ireland. *Norsk geol. Tidsskr.* **50,** 19–44.

DIGGENS, J. N. & ROMANO, M. 1968. The Caradoc rocks around Llyn Cowlyd, North Wales. *Geol. J.* **6,** 31–48. [32].

DIXON, E. E. L. 1921. The geology of the South Wales Coalfield. Part XIII. The country around Pembroke and Tenby. (Sheets 244 & 245). *Mem. geol. Surv. U.K.*, i–vi, 1–220. [113].

—— 1925. Field work: Whitehaven Sheet. *Summ. Progr. geol. Surv. Lond.* (for 1924), 70–1.

DOWNIE, C. & FORD, T. D. 1966. Microfossils from the Manx Slate Series. *Proc. Yorks. geol. Soc.* **35,** 307–22.

——, LISTER, T. R., HARRIS, A. L. & FETTES, D. J. 1971. A palynological investigation of the Dalradian rocks of Scotland. *Rep. No.* 71/9, *Inst. geol. Sci.*, i–iv, 1–29.

DREW, H. & SLATER, I. L. 1910. Notes on the geology of the district around Llansawel (Carmarthenshire). *Q. Jl geol. Soc. Lond.* **66,** 402–19. [83].

EASTWOOD, T., DIXON, E. E. L., HOLLINGWORTH, S. E. & SMITH, B. 1931. The geology of the Whitehaven and Workington district. (Sheet 28). *Mem. geol. Surv. U.K.*, i–xvii, 1–304.

——, HOLLINGWORTH, S. E., ROSE, W. C. C. & TROTTER, F. M. 1968. Geology of the country around Cockermouth and Caldbeck. (Sheet 23). *Mem. geol. Surv. U.K.*, i–x, 1–298.

ECKFORD, R. J. A. & RITCHIE, M. 1931. The lavas of Tweeddale and their position in the Caradocian sequence. *Summ. Progr. geol. Surv. Lond.* (for 1930), Pt. 3, 46–57.

EKSTRÖM, G. 1937. Upper Didymograptus Shale in Scania. *Sver. geol. Unders. Afh.* Ser. C. **403,** 1–53.

ELLES, G. L. 1904. Some graptolite zones in the Arenig rocks of Wales. *Geol. Mag.* **41,** 199–211. [24].

—— 1909. The relation of the Ordovician and Silurian rocks of Conway (North Wales). *Q. Jl geol. Soc. Lond.* **65,** 169–94. [35].

—— 1922a. The Bala country: its structure and rock succession. *Q. Jl geol. Soc. Lond.* **78,** 132–75. [49].

—— 1922b. A new *Azygograptus* from North Wales. *Geol. Mag.* **59,** 299–301.

—— 1925. The characteristic assemblages of the graptolite zones of the British Isles. *Geol. Mag.* **62,** 337–47.

—— 1933. The Lower Ordovician graptolite faunas with special reference to the Skiddaw Slates. *Summ. Progr. geol. Surv. Lond.* (for 1932), Pt. 2, 94–111.

—— 1940. The stratigraphy and faunal succession in the Ordovician rocks of the Builth-Llandrindod Inlier, Radnorshire. *Q. Jl geol. Soc. Lond.* **95** (for 1939), 383–445. [95].

—— & WOOD, E. M. R. 1901–1918. A monograph of British graptolites. *Palaeontogr. Soc.* (*Monogr.*), *a–m*, i–clxxi, 1–530.

ERDTMANN, B.-D. 1965. Outline stratigraphy of graptolite-bearing 3b (Lower Ordovician) strata in the Oslo region, Norway. *Norsk geol. Tidsskr.* **45,** 481–547.

EVANS, C. D. R. 1968. *Geological succession and structure of the area east of Bethesda.* Thesis (Ph.D.) Wales. [27].

EVANS, D. C. 1906. The Ordovician rocks of western Caermarthenshire. *Q. Jl geol. Soc. Lond.* **62,** 597–643. [98].

EVANS, J. W. & STUBBLEFIELD, C. J. [Eds.] 1929. *Handbook of the geology of Great Britain: a compilative work.* Murby, London, i–xii, 1–556.

EVANS, W. D. 1945. The geology of the Prescelly Hills, north Pembrokeshire. *Q. Jl geol. Soc. Lond.* **101,** 89–110. [99].

—— 1948. The Cambrian-Ordovician junction, Whitesand Bay, Pembrokeshire. *Geol. Mag.* **85,** 110–2.

FEARNSIDES, W. G. 1905. On the geology of Arenig Fawr and Moel Llyfnant. *Q. Jl geol. Soc. Lond.* **61,** 608–40. [37].

—— 1910. The Tremadoc Slates and associated rocks of south-east Carnarvonshire. *Q. Jl geol. Soc. Lond.* **66,** 142–88. [14].

—— & DAVIES, W. 1944. The geology of Deudraeth. The country between Traeth Mawr and Traeth Bâch, Merioneth. *Q. Jl geol. Soc. Lond.* **99** (for 1943), 247–76. [15].

——, ELLES, G. L. & SMITH, B. 1907. The Lower Palaeozoic rocks of Pomeroy. *Proc. R. Ir. Acad.* Sec. B. **26,** 97–128.

FIRMAN, R. J. 1957. The Borrowdale Volcanic Series between Wastwater and Duddon Valley, Cumberland. *Proc. Yorks. geol. Soc.* **31,** 39–64.

FITCH, F. J. 1967. Ignimbrite volcanism in North Wales. *Bull. volcan.* **30,** 199–219. [6].

FRANCIS, E. H., HOWELLS, M. F., LEVERIDGE, B. E. & EVANS, C. D. R. 1970. Current work by the Institute of Geological Sciences in North Wales. *Proc. geol. Soc. Lond.* **1663,** 165–7.

GARDINER, C. I. & REYNOLDS, S. H. 1897. An account of the Portraine inlier (Co. Dublin) with an appendix on the fossils by F. R. C. Reed. *Q. Jl geol. Soc. Lond.* **53,** 520–39.

—— & —— 1909. On the igneous and associated sedimentary rocks of the Tourmakeady district (County Mayo); with a palaeontological appendix by F. R. C. Reed. *Q. Jl geol. Soc. Lond.* **65,** 104–54.

—— & —— 1910. The igneous and sedimentary rocks of the Glensaul district (County Galway). With a Palaeontological appendix by F. R. C. Reed. *Q. Jl geol. Soc. Lond.* **66,** 253–80.

—— & —— 1912. The Ordovician and Silurian rocks of the Kilbride peninsula (Mayo). *Q. Jl geol. Soc. Lond.* **68,** 75–102.

—— & —— 1914. The Ordovician and Silurian rocks of the Lough Nafooey area (County Galway). *Q. Jl geol. Soc. Lond.* **70,** 104–18.

GARDINER, P. R. R. 1967. *The geology of the Lower Palaeozoic rocks in the Duncannon area, Co. Wexford.* Thesis (Ph.D.) Dublin.

—— & BRENCHLEY, P. J. 1970. The Pre-Cambrian and Lower Palaeozoic geology of Co. Wexford. *Ir. Nat. J.* **16,** 371–9.

GEORGE, T. N. 1970. *British Regional Geology. South Wales.* Third edition. Institute of Geological Sciences. H.M.S.O., London, i–xii, 1–152.

GOBBETT, D. J. & WILSON, C. B. 1960. The Oslobreen Series, Upper Hecla Hoek of Ny Friesland, Spitsbergen. *Geol. Mag.* **97,** 441–60.

GREEN, J. F. N. 1908. The geological structure of the St. David's area (Pembrokeshire). *Q. Jl geol. Soc. Lond.* **64,** 363–83. [107].

—— 1915. The structure of the eastern part of the Lake District. *Proc. Geol. Ass.* **26,** 195–233.

GREENLY, E. 1919. The geology of Anglesey. *Mem. geol. Surv. U.K.,* i–xl, 1–980. In 2 vols.

—— 1944. The Ordovician rocks of Arvon. *Q. Jl geol. Soc. Lond.* **100,** 75–83. [25].

GREIG, D. C., WRIGHT, J. E., HAINS, B. A. & MITCHELL, G. H. 1968. Geology of the country around Church Stretton, Craven Arms, Wenlock Edge and Brown Clee. (Sheet 166). *Mem. geol. Surv. U.K.,* i–xiv, 1–380. [73].

GROOM, T. T. 1902. The sequence of the Cambrian and associated beds of the Malvern Hills. *Q. Jl geol. Soc. Lond.* **58,** 89–149.

—— & LAKE, P. 1908. The Bala and Llandovery rocks of Glyn Ceiriog (North Wales). *Q. Jl geol. Soc. Lond.* **64,** 546–95. [54].

HAINS, B. A. 1969. *The geology of the Craven Arms area. (Explanation of 1:25 000 Geological Sheet SO48).* Institute of Geological Sciences. H.M.S.O., London, i–vi, 1–70. [67].

—— 1970. *The geology of the Wenlock Edge area. (Explanation of 1:25 000 Geological Sheet SO59).* Institute of Geological Sciences. H.M.S.O., London, i–vi, 1–61. [69].

HALL, T. S. 1899. The graptolite-bearing rocks of Victoria, Australia. *Geol. Mag.* **36,** 438–51.

HARKER, A. 1891. The ancient lavas of the Lake District. *Naturalist,* 145–7.

HARKNESS, R. & NICHOLSON, H. A. 1877. On the strata and their fossil contents between the Borrowdale Series of the north of England and the Coniston Flags. *Q. Jl geol. Soc. Lond.* **33,** 461–84.

HARPER, J. C. 1947. The Caradoc fauna of Ynys Galed, Caernarvonshire. *Ann. Mag. nat. Hist.* Ser. 11. **14,** 153–75.

—— 1948. The Ordovician and Silurian rocks of Ireland. *Proc. Lpool geol. Soc.* **20,** 48–67.

—— 1952. The Ordovician rocks between Collon (Co. Louth) and Grangegeeth (Co. Meath). *Scient. Proc. R. Dubl. Soc.* **26** n.s., 85–112.

—— 1956. The Ordovician succession near Llanystwmdwy, Caernarvonshire. *Lpool Manchr geol. J.* **1,** Pt. 4 (1954), 385–93. [12].

—— & RAST, N. 1964. The faunal succession and volcanic rocks of the Ordovician near Bellewstown, Co. Meath. *Proc. R. Ir. Acad. Sec. B.* **64,** 1–23.

HARRIS, W. J. & THOMAS, D. E. 1938. A revised classification and correlation of the Ordovician graptolite beds of Victoria. *Min. geol. J.,* **1,** (3), 62–72.

HARTLEY, J. J. 1925. The succession and structure of the Borrowdale Volcanic Series as developed in the area lying between the lakes of Grasmere, Windermere and Coniston. *Proc. Geol. Ass.* **36,** 203–26.

—— 1932. The volcanic and other igneous rocks of Great and Little Langdale. *Proc. Geol. Ass.* **43,** 32–69.

—— 1933. The geology of north-east Tyrone and the adjacent portions of Co. Londonderry. *Proc. R. Ir. Acad. Sec. B.* **41,** 218–55.

—— 1936. The age of the igneous series of Slieve Gallion, Northern Ireland. *Geol. Mag.* **73,** 226–8.

—— 1942. The geology of Helvellyn and the southern part of Thirlmere. *Q. Jl geol. Soc. Lond.* **97,** 129–62.

—— & HARPER, J. C. 1937. A recently discovered Ordovician inlier in Co. Down, with a note on a new species of *Pyritonema. Ir. Nat. J.* **6,** 253–5.

HAVLÍČEK, V. & VANĚK, J. 1966. The biostratigraphy of the Ordovician of Bohemia. *Sb. geol. Ved. Praha. Rada. P.* **8,** 7–69.

HAWKINS, T. R. W. 1966. Boreholes at Parys mountain, near Amlwch, Anglesey. *Bull. geol. Surv. Gt. Br.* No. 24, 7–18.

HELM, D. G. 1970. Stratigraphy and structure in the Black Combe Inlier, English Lake District. *Proc. Yorks. geol. Soc.* **38,** 105–48.

HENDRICKS, E. M. L. 1926. The Bala-Silurian succession in the Llangranog district (south Cardiganshire). *Geol. Mag.* **63,** 121–39. [86].

HICKS, H. 1875. On the succession of the ancient rocks in the vicinity of St. David's, Pembrokeshire, with special reference to those of the Arenig and Llandeilo Groups, and their fossil contents. *Q. Jl geol. Soc. Lond.* **31**, 167–95.

—— 1881. The classification of the Eozoic and Lower Palaeozoic rocks of the British Isles. *Pop. Sci. Rev.* **5**, 289–308.

HIGGINS, A. C. 1967. The age of the Durine member of the Durness Limestone Formation at Durness. *Scott. J. Geol.* **3**, 382–8.

HOLLAND, C. H. 1971. Cambrian-Ordovician boundary in the British Isles. *Proc. geol. Soc. Lond.* **1664** (for 1970), 229–30.

HOLLINGWORTH, S. E. 1954. The geology of the Lake District—a review. *Proc. Geol. Ass.* **65**, 385–401.

HOPKINSON, J. & LAPWORTH, C. 1875. Descriptions of the graptolites of the Arenig and Llandeilo rocks of St. David's. *Q. Jl geol. Soc. Lond.* **31**, 631–72.

HUGHES, C. P. 1969. The Ordovician trilobite faunas of the Builth-Llandrindod Inlier, central Wales. Part 1. *Bull. Br. Mus. nat. Hist.* (Geol.) **18**, 39–103.

HUGHES, E. W. 1917. On the geology of the district from Cil-y-Coed to the St. Annes-Llanllyfni Ridge (Carnarvonshire). *Geol. Mag.* **54**, 12–25. [23].

INGHAM, J. K. 1966. The Ordovician rocks in the Cautley and Dent districts of Westmorland and Yorkshire. *Proc. Yorks. geol. Soc.* **35**, 455–505.

—— 1968. British and Swedish Ordovician species of *Cybeloides* (Trilobita). *Scott. J. Geol.* **4**, 300–15.

—— 1970. A monograph of the Upper Ordovician trilobites from the Cautley and Dent districts of Westmorland and Yorkshire. *Palaeontogr. Soc.* (*Monogr.*) Pt. 1, 1–58.

—— & WRIGHT, A. D. 1970. A revised classification of the Ashgill Series. *Lethaia* **3**, 233–42.

JAANUSSON, V. 1960. Graptoloids from the Ontikan and Viruan (Ordov.) Limestones of Estonia and Sweden. *Bull. geol. Inst. Upps.* **38**, 289–366.

—— & STRACHAN, I. 1954. Correlation of the Scandinavian Middle Ordovician with the graptolite succession. *Geol. För. Stockh. Förh.* **76**, 684–96.

JACKSON, D. E. 1961. Stratigraphy of the Skiddaw Group between Buttermere and Mungrisdale, Cumberland. *Geol. Mag.* **98**, 515–28.

—— 1962. Graptolite zones in the Skiddaw Group in Cumberland, England. *J. Paleont.* **36**, 300–13.

JAMES, D. M. D. 1968. *Sedimentary studies in the Bala of central Wales.* Thesis (Ph.D.) Wales.

—— 1971a. The Nant-y-moch Formation, Plynlimon Inlier, west central Wales. *Jl geol. Soc. Lond.* **127**, 177–81. [76].

—— 1971b. Petrography of the Plynlimon Group, west central Wales. *Sedim. Geol.* **6**, 225–70.

—— & JAMES, J. 1969. The influence of deep fractures on some areas of Ashgillian-Llandoverian sedimentation in Wales. *Geol. Mag.* **106**, 562–82.

JEHU, R. M. 1926. The geology of the district around Towyn and Abergynolwyn (Merioneth). *Q. Jl geol. Soc. Lond.* **82**, 465–89. [43].

JEHU, T. J. & CAMPBELL, R. 1917. The Highland Border rocks of the Aberfoyle district. *Trans. R. Soc. Edinb.* **52**, 175–212.

JENNINGS, A. V. & WILLIAMS, G. J. 1891. On Manod and the Moelwyns. *Q. Jl geol. Soc. Lond.* **47**, 368–83. [16].

JOHNSON, G. A. L. 1961. Skiddaw Slates proved in the Teesdale Inlier. *Nature, Lond.* **190**, 996–7.

JONES, B. 1933. The geology of the Fairbourne-Llwyngwril district, Merioneth. *Q. Jl geol. Soc. Lond.* **89**, 145–71.

JONES, G. H. 1956. *The geology of the western Berwyns.* Thesis (Ph.D.) Birmingham. [57].

JONES, O. T. 1909. The Hartfell-Valentian succession in the district around Plynlimon and Pont Erwyd (north Cardiganshire). *Q. Jl geol. Soc. Lond.* **65**, 463–536.

—— 1922. Lead and zinc. The mining district of north Cardiganshire and west Montgomeryshire. *Mem. geol. Surv. spec. Rep. Miner. Resour. Gt. Br.* **20**, i–vi, 1–207. [74].

—— 1925. The geology of the Llandovery district: Part I.—The southern area. *Q. Jl geol. Soc. Lond.* **81**, 344–88. [89].

—— 1936. The Lower Palaeozoic rocks of Britain. 16th *Int. geol. Congr.*, Washington, 1933, **1**, 463–84.

—— 1938. On the evolution of a geosyncline. *Q. Jl geol. Soc. Lond.* **94**, lx–cx.

—— 1940. Some Lower Palaeozoic contacts in Pembrokeshire. *Geol. Mag.* **77**, 405–9.

—— 1947. The geology of the Silurian rocks west and south of the Carneddau Range, Radnorshire. *Q. Jl geol. Soc. Lond.* **103**, 1–36. [93].

—— 1949. The geology of the Llandovery district. Part II. The northern area. *Q. Jl geol. Soc. Lond.* **105**, 43–64. [90].

—— 1954. The use of graptolites in geological mapping. *Lpool Manchr geol. J.* **1** (for 1952–54), 246–60.

—— 1956. The geological evolution of Wales and the adjacent regions. *Q. Jl geol. Soc. Lond.* **111** (for 1955), 323–51.

—— & Pugh, W. J. 1916. The geology of the district around Machynlleth and the Llyfnant valley. *Q. Jl geol. Soc. Lond.* **71** (for 1915), 343–85. [44].

—— & —— 1935. The geology of the districts around Machynlleth and Aberystwyth. *Proc. Geol. Ass.* **46**, 247–300.

—— & —— 1941. The Ordovician rocks of the Builth district; a preliminary account. *Geol. Mag.* **78**, 185–91.

—— & —— 1946. The complex intrusion of Welfield, near Builth Wells, Radnorshire. *Q. Jl geol. Soc. Lond.* **102**, 157–88.

—— & —— 1948. A multi-layered dolerite complex of laccolithic form, near Llandrindod Wells, Radnorshire. *Q. Jl geol. Soc. Lond.* **104**, 43–70.

—— & —— 1949. An early Ordovician shore-line in Radnorshire, near Builth Wells. *Q. Jl geol. Soc. Lond.* **105**, 65–99. [94].

Jones, W. D. V. 1945. The Valentian succession around Llanidloes, Montgomeryshire. *Q. Jl geol. Soc. Lond.* **100** (for 1944), 309–32. [77].

Kaljo, D. & Klaamann, E. 1965. The fauna of the Portrane Limestone. III. The corals. *Bull. Br. Mus. nat. Hist.* (Geol.) **10**, 413–34.

Keeping, W. 1881. The geology of Central Wales. With an Appendix on some new species of Cladophora, by Charles Lapworth. *Q. Jl geol. Soc. Lond.* **37**, 141–77.

—— 1882. On the geology of Cardigan town. *Geol. Mag.* **19**, 519–22.

Kelling, G. 1961. The stratigraphy and structure of the Ordovician rocks of the Rhinns of Galloway. *Q. Jl geol. Soc. Lond.* **117**, 37–75.

Kielan, Z. 1960 [dated 1959]. Upper Ordovician trilobites from Poland and some related forms from Bohemia and Scandinavia. *Palaeont. pol.* **11**, i–vi, 1–198.

Kilroe, J. R. 1907. The Silurian and metamorphic rocks of Mayo and north Galway. *Proc. R. Ir. Acad. Sec. B.* **26**, 129–60.

King, W. B. R. 1923. The Upper Ordovician rocks of the south-western Berwyn Hills. *Q. Jl geol. Soc. Lond.* **79**, 487–507. [59].

—— 1928. The geology of the district around Meifod (Montgomeryshire). *Q. Jl geol. Soc. Lond.* **84**, 671–702. [62].

—— 1932. A fossiliferous limestone associated with Ingletonian beds at Horton-in-Ribblesdale, Yorkshire. *Q. Jl geol. Soc. Lond.* **88**, 100–11.

—— & Wilcockson, W. H. 1934. The Lower Palaeozoic rocks of Austwick and Horton-in-Ribblesdale, Yorkshire. *Q. Jl geol. Soc. Lond.* **90**, 7–31.

—— & Williams, A. 1948. On the lower part of the Ashgillian Series in the north of England. *Geol. Mag.* **85**, 205–12.

Lamont, A. 1935. The Drummuck Group, Girvan; a stratigraphical revision, with descriptions of new fossils from the lower part of the group. *Trans. geol. Soc. Glasg.* **19**, 288–334.

—— 1941. Trinucleidae in Eire. *Ann. Mag. nat. Hist.* Ser. II. **8**, 438–69.

—— 1946. B. B. Bancroft's work on Caradocian lavas and sediments in the Capel Curig district, Caernarvonshire.—I. *Quarry Mgrs J.* **30**, 235–9.

—— & Lindström, M. 1957. Arenigian and Llandeilian cherts identified in the Southern Uplands of Scotland by conodonts, etc. *Trans. Edinb. geol. Soc.* **17**, 60–70.

LAPWORTH, C. 1876–77. On the graptolites of County Down. *Rep.Proc.Belf.Nat. Fld Club*. Ser. 2. 1, appendix IV, 125–44.

—— 1879a. On the tripartite classification of the Lower Palaeozoic rocks. *Geol. Mag.* **16**, 1–15.

—— 1879b. On the geological distribution of the Rhabdophora. *Ann. Mag. nat. Hist.* Ser. 5. **3**, 245–57, 449–55; **4**, 333–41, 423–31.

—— 1880. On the geological distribution of the Rhabdophora [cont.]. *Ann. Mag. nat. Hist.* Ser. 5, **5**, 45–62, 273–85, 358–69; **6**, 16–29, 185–207.

—— & WATTS, W. W. 1894. The geology of south Shropshire, with special reference to the district to be visited during the long excursion. *Proc. Geol. Ass.* **13**, 297–355.

LAPWORTH, H. 1900. The Silurian sequence of Rhayader. *Q. Jl geol. Soc. Lond.* **56**, 67–137. [79].

LEEDAL, G. P. & WALKER, G. P. L. 1950. A restudy of the Ingletonian Series of Yorkshire. *Geol. Mag.* **87**, 57–66.

LEWIS, H. P. 1926. On *Bolopora undosa* gen. et sp. nov.: a rock building bryozoan with phosphatized skeleton, from the basal Arenig rocks of Ffestiniog (North Wales). [With a stratigraphical note on the Garth Grit by W. G. Fearnsides.] *Q. Jl geol. Soc. Lond.* **82**, 411–27.

LINDSTRÖM, M. 1959. Conodonts from the Crug Limestone (Ordovician, Wales). *Micropaleontology* **5**, 427–52.

LINNARSSON, J. G. O. 1876. A comparison between the oldest fossiliferous rocks of northern Europe. *Geol. Mag.* **13**, 145–50.

LISTER, T. R., COCKS, L. R. M. & RUSHTON, A. W. A. 1969. The basement beds in the Bobbing borehole, Kent. *Geol. Mag.* **106**, 601–3.

LYNAS, B. D. T. 1970. *The geology of the Migneint area, North Wales*. Thesis (Ph.D.) Cambridge. [36].

MACGREGOR, A. R. 1958. *The Llandeilo Limestone of the central Berwyns and a comparison with rocks of the same age in north-western Europe*. Thesis (Ph.D.) Cambridge. [58].

—— 1961. Upper Llandeilo brachiopods from the Berwyn Hills, North Wales. *Palaeontology* **4**, 177–209.

—— 1963. Upper Llandeilo trilobites from the Berwyn Hills, North Wales. *Palaeontology* **5**, 790–816.

McKERROW, W. S. & CAMPBELL, C. J. 1960. The stratigraphy and structure of the Lower Palaeozoic rocks of north-west Galway. *Scient. Proc. R. Dubl. Soc.* Ser. A. **1**, 27–51.

MÄNNIL, R. 1966. *Evolution of the Baltic Basin during the Ordovician*. Tallin, 1–199. (In Russian, English summary).

—— 1971. Distribution of selected Ordovician chitinozoan assemblages and species in northern Europe and their stratigraphical evaluation. *Mém. Bur. Rech. géol. minièr.* **73**, 309–11.

MARR, J. E. 1892. The Coniston Limestone Series. *Geol. Mag.* **29**, 97–110.

—— 1905. [The classification of the sedimentary rocks.] *Q. Jl geol. Soc. Lond.* **61**, lxi–lxxxvi.

—— 1906. On the stratigraphical relations of the Dufton Shales and the Keisley Limestone of the Cross Fell Inlier. *Geol. Mag.* **43**, 481–7.

—— 1907. The Ashgillian Series. *Geol. Mag.* **44**, 59–69.

—— 1913. The Lower Palaeozoic rocks of the Cautley district. *Q. Jl geol. Soc. Lond.* **69**, 1–17.

—— 1916. The Ashgillian succession to the west of Coniston Lake. *Q. Jl geol. Soc. Lond.* **71**, 189–204.

—— & ROBERTS, T. 1885. The Lower Palaeozoic rocks of the neighbourhood of Haverfordwest. *Q. Jl geol. Soc. Lond.* **41**, 476–91.

MATLEY, C. A. 1928. The Pre-Cambrian complex and associated rocks of south-western Lleyn (Carnarvonshire), with a chapter on the petrology of the complex by E. Greenly. *Q. Jl geol. Soc. Lond.* **84**, 440–504. [3].

—— 1932. The geology of the country around Mynydd Rhiw and Sarn, south-western Lleyn. *Q. Jl geol. Soc. Lond.* **88**, 238–73. [4].

—— 1938. The geology of the country around Pwllheli, Llanbedrog and Madryn, south-west Carnarvonshire. *Q. Jl geol. Soc. Lond.* **94**, 555–606. [7].

—— & HEARD, A. 1930. The geology of the country around Bodfean (south-western Carnarvonshire). *Q. Jl geol. Soc. Lond.* **86**, 130–68. [8].

—— & SMITH, B. 1936. The age of the Sarn granite. *Q. Jl geol. Soc. Lond.* **92**, 188–200.

MITCHELL, G. H. 1925. The Coniston Limestone Series of the Kentmere district. *Geol. Mag.* **62**, 264-7.

—— 1929. The succession and structure of the Borrowdale Volcanic Series in Troutbeck, Kentmere and the western part of Long Sleddale (Westmorland). *Q. Jl geol. Soc. Lond.* **85**, 9-44.

—— 1934. The Borrowdale Volcanic Series and associated rocks in the country between Long Sleddale and Shap. *Q. Jl geol. Soc. Lond.* **90**, 418-44.

—— 1940. The Borrowdale Volcanic Series of Coniston, Lancashire. *Q. Jl geol. Soc. Lond.* **96**, 301-19.

—— 1956a. The geological history of the Lake District. *Proc. Yorks. geol. Soc.* **30**, 407-63.

—— 1956b. The Borrowdale Volcanic Series of the Dunnerdale Fells, Lancashire. *Lpool Manchr geol. J.* **1**, 428-49.

—— 1963. The Borrowdale volcanic rocks of the Scathwaite Fells, Lancashire. *Lpool Manchr geol. J.* **3**, 289-300.

MITCHELL, W. I. 1969. *The Lower Palaeozoic rocks of Pomeroy, Co. Tyrone.* Thesis (B.Sc.) Belfast.

——, CARLISLE, H., HILLIER, N. & ADDISON, R. 1972. A correlation of the Ordovician rocks of Courtown (Co. Wexford) and Tramore (Co. Waterford). *Proc. R. Ir. Acad. Sec. B.* **72**, 83-9.

MONSEN, A. 1937. Die Graptolithenfauna im Unteren Didymograptusschiefer (Phyllograptus-schiefer) Norwegens. *Norsk geol. Tidsskr.* **16** (for 1936), 57-266.

MOSELEY, F. 1960. The succession and structure in the Borrowdale volcanic rocks south east of Ullswater. *Q. Jl geol. Soc. Lond.* **116**, 55-84.

MURCHISON, R. I. 1834. On the structure and classification of the Transition rocks of Shropshire, Herefordshire, and part of Wales, and on the lines of disturbance which have affected that series of deposits, including the Valley of Elevation of Woolhope. *Proc. geol. Soc. Lond.* **2** (for 1833-38), 13-18.

—— 1835. On the Silurian System of rocks. *Lond. Edinb. Phil. Mag.* Ser. 3. **7**, 46-52.

—— 1839. *The Silurian System founded on geological researches in the counties of Salop, Hereford, Radnor, Montgomery, Caermarthen, Brecon, Pembroke, Monmouth, Gloucester, Worcester, and Stafford; with descriptions of the coal-fields and overlying formations.* Murray, London, i-xxxii, 1-768. (In 2 vols.).

—— 1854. *Siluria. The history of the oldest known rocks containing organic remains, with a brief sketch of the distribution of gold over the earth.* Murray, London, i-xvi, 1-523. (2nd ed. 1859, 3rd ed. 1867, 4th ed. 1872).

MURPHY, G. J. 1958. [Excursion to Tramore Bay]. *In* Brindley, J. C. & GILL, W. D. Summer field meeting in southern Ireland, 1957. *Proc. Geol. Ass.* **69**, 255-7.

NICHOLAS, T. C. 1915. The geology of the St. Tudwal's peninsula (Carnarvonshire). *Q. Jl geol. Soc. Lond.* **71**, 83-143. [5].

NICHOLSON, H. A. & MARR, J. E. 1887. On the occurrence of a new fossiliferous horizon in the Ordovician series of the Lake District. *Geol. Mag.* **24**, 339-44.

—— & —— 1891. The Cross Fell Inlier. *Q. Jl geol. Soc. Lond.* **47**, 500-12.

NIKITIN, I. F. 1971. The Ordovician System in Kazakhstan. *Mém. Bur. Rech. géol. minièr.* **73**, 337-43.

——, APOLLONOV, M. K. & TZAY, D. T. 1968. Correlation scheme for the Ordovician of eastern Kazakhstan. *Izv. Akad. Nauk. kazakh. S.S.R.* Ser. Geol. **3**, 1-12. (In Russian).

OLIVER, R. L. 1954. Welded tuffs in the Borrowdale Volcanic Series, English Lake District, with a note on similar rocks in Wales. *Geol. Mag.* **91**, 473-83.

O'NIONS, R. K., OXBURGH, E. R., HAWKSWORTH, C. J. & MACINTYRE, R. M. 1972. New isotopic and stratigraphic evidence on the age of the Ingletonian. *Jl geol. Soc. Lond.* (in press).

ÖPIK, A. A. 1958. The geology of the Canberra City district. *Bull Bur. Miner. Resour. Geol. Geophys. Aust.* **32**, 1-99.

PEACH, B. N., HORNE, J., GUNN, W., CLOUGH, C. T. & HINXMAN, L. W. 1907. The geological structure of the North-west Highlands of Scotland. *Mem. geol. Surv. U.K.*, i-xviii, 1-668.

PHILLIPS, W. E. A. & SKEVINGTON, D. 1968. The Lower Palaeozoic rocks of the Lough Acanon area, Co. Cavan, Ireland. *Scient. Proc. R. Dubl. Soc.* **3**, 141-8.

POCOCK, R. W., WHITEHEAD, T. H., WEDD, C. B. & ROBERTSON, T. 1938. Shrewsbury district, including the Hanwood Coalfield. (Sheet 152). *Mem. geol. Surv. U.K.*, i-xxi, 1-297. [73].

POTTER, J. F. 1960. *Ludlovian-Downtonian rocks in the neighbourhood of Llandeilo, Carmarthenshire.* Thesis (Ph.D.) London. [88].

PRICE, D. 1971. *The stratigraphy and trilobite fauna of the Sholeshook Limestone of South Wales.* Thesis (Ph.D.) London.

PRINGLE, J. 1930. The geology of Ramsey Island (Pembrokeshire). *Proc. Geol. Ass.* **41**, 1–31. [108].

—— 1961. *British Regional Geology. The south of Scotland.* Geological Survey and Museum. H.M.S.O., Edinburgh, i–iv, 1–87.

PUGH, W. J. 1923. The geology of the district around Corris and Aberllefenni (Merionethshire). *Q. Jl geol. Soc. Lond.* **79**, 508–45. [45].

—— 1928. The geology of the district around Dinas Mawddwy (Merioneth). *Q. Jl geol. Soc. Lond.* **84**, 345–81. [46].

—— 1929. The geology of the district between Llanymawddwy and Llanuwchllyn (Merioneth). *Q. Jl geol. Soc. Lond.* **85**, 242–306. [47].

PULFREY, W. 1933. The iron-ore oolites and pisolites of North Wales. *Q. Jl geol. Soc. Lond.* **89**, 401–30.

RAMSAY, J. G. 1959. Cwm Idwal-Capel Curig. Itinerary II of: Williams, D. & Ramsay, J. G. Geology of some classic British areas: Snowdonia. *Geologists' Association Guides* No. 28, 10–4. [29].

RAST, N. 1961. Mid-Ordovician structures in south-western Snowdonia. *Lpool Manchr geol. J.* **2**, 645–52. [19].

—— 1969. The relationship between Ordovician structure and volcanicity in Wales. *In* Wood, A. (Ed.) *The Pre-Cambrian and Lower Palaeozoic rocks of Wales.* University of Wales Press, Cardiff, 305–35.

——, BEAVON, R. V. & FITCH, F. J. 1958. Sub-aerial volcanicity in Snowdonia. *Nature, Lond.* **181**, 508.

RASTALL, R. H. 1910. The Skiddaw Granite and its metamorphism. *Q. Jl geol. Soc. Lond.* **66**, 116–40.

REED, F. R. C. 1895. The geology of the country around Fishguard. *Q. Jl geol. Soc. Lond.* **51**, 149–95. [100].

—— 1896. The fauna of the Keisley Limestone, Part i. *Q. Jl geol. Soc. Lond.* **52**, 407–37.

—— 1897. The fauna of the Keisley Limestone, Part ii. *Q. Jl geol. Soc. Lond.* **53**, 67–106.

—— 1899. The Lower Palaeozoic bedded rocks of County Waterford. *Q. Jl geol. Soc. Lond.* **55**, 718–72.

—— 1952. Revision of certain Ordovician fossils from County Tyrone. *Proc. R. Ir. Acad.* Sec. B. **55**, 29–136.

REYNOLDS, S. H. & GARDINER, C. I. 1896. The Kildare Inlier. *Q. Jl geol. Soc. Lond.* **52**, 587–605.

RHODES, F. H. T. 1953. Some British Lower Palaeozoic conodont faunas. *Phil. Trans. R. Soc.* Ser. B. **237** (for 1952–54), 261–334.

ROBERTS, B. 1967. Succession and structure in the Llwyd Mawr syncline, Caernarvonshire, North Wales. *Geol. J.* **5**, 369–90. [13].

ROBERTS, R. O. 1927. The igneous and associated Ordovician rocks of Baxter's Bank, Radnorshire. *Geol. Mag.* **64**, 289–98.

—— 1929. The geology of the district around Abbey-Cwmhir (Radnorshire). *Q. Jl geol. Soc. Lond.* **85**, 651–76. [78].

ROMANO, M. & DIGGENS, J. N. 1969. Longvillian shelly faunas from the Dolwyddelan area, North Wales. *Geol. Mag.* **106**, 603–6 [Corr.].

RÖÖMUSOKS, A. 1960. Stratigraphy and paleogeography of the Ordovician in Estonia. 21st *Int. geol. Congr.*, Copenhagen, **7**, 58–69.

—— 1970. *Stratigraphy of the Viruan Series (Middle Ordovician) in northern Estonia.* Tallin, 1–343. (In Russian).

ROSE, W. C. C. 1954. The sequence and structure of the Skiddaw Slates in the Keswick-Buttermere area. *Proc. Geol. Ass.* **65**, 403–6.

ROSS, J. R. P. P. 1966. The fauna of the Portrane Limestone, IV: Polyzoa. *Bull. Br. Mus. nat. Hist.* (Geol.) **12**, 107–35.

ROSS, R. J. & INGHAM, J. K. 1970. Distribution of the Toquima-Table Head (Middle Ordovician Whiterock) faunal realm in the Northern Hemisphere. *Bull. geol. Soc. Am.* **81**, 393–408.

SARJENT, H. C. 1924. Notes on the petrology of Penmaenmawr Mountain. (Part I). *Proc. Lpool geol. Soc.* **14** (for 1923–27), (1), 82–98.

SCHIENER, E. J. 1970. Sedimentology and petrography of three tuff horizons in the Caradocian sequence of the Bala area (North Wales). *Geol. J.* **7**, 25–46.

SEDGWICK, A. 1938. A synopsis of the English series of stratified rocks inferior to the old red sandstone—with an attempt to determine the successive natural groups and formations. *Proc. Geol. Soc. Lond.* **2** (for 1833–38), 675–85.

—— 1845. On the older Palaeozoic (Protozoic) rocks of North Wales. *Q. Jl geol. Soc. Lond.* **1**, 5–22.

—— 1852. [Advertisement to the second fascicule.] *In* Sedgwick, A. & M'Coy, F. *A synopsis of the classification of the British Palaeozoic rocks. With a systematic description of the British Palaeozoic fossils in the Geological Museum of the University of Cambridge.* In 2 vols. Parker, London, iii–viii.

SHACKLETON, R. M. 1954. The structural evolution of North Wales. *Lpool Manchr geol. J.* **1**, Pt. 3 (for 1953), 261–97.

—— 1959. The stratigraphy of the Moel Hebog district between Snowdon and Tremadoc. With a note on the Caradoc shelly faunas by J. C. Harper. *Lpool Manchr geol. J.* **2**, 216–52. [20].

SHARPE, E. N. 1970. An occurrence of pillow lavas in the Ordovician of Co. Down. *Ir. Nat. J.* **16**, 299–301.

SHIRLEY, J. 1936. Some British trilobites of the family Calymenidae. *Q. Jl geol. Soc. Lond.* **92**, 384–422.

SHOTTON, F. W. 1935. The stratigraphy and tectonics of the Cross Fell Inlier. *Q. Jl geol. Soc. Lond.* **91**, 639–704.

SIMPSON, A. 1963. The stratigraphy and tectonics of the Manx Slate Series, Isle of Man. *Q. Jl geol. Soc. Lond.* **119**, 367–400.

—— 1967. The structure and tectonics of the Skiddaw Slates and the relationship of the overlying Borrowdale Volcanic Series in part of the Lake District. *Geol. J.* **5**, 391–418.

SKEVINGTON, D. 1968. British and North American Lower Ordovician correlation: discussion. *Bull. geol. Soc. Am.* **79**, 1259–64.

—— 1969. The classification of the Ordovician System in Wales. *In* Wood, A. (Ed.) *The Pre-Cambrian and Lower Palaeozoic rocks of Wales.* University of Wales Press, Cardiff, 161–79.

—— 1970a. A Lower Llanvirn graptolite fauna from the Skiddaw Slates, Westmorland. *Proc. Yorks. geol. Soc.* **37**, 395–444.

—— 1970b. Graptolite faunal provinces in Ordovician of northwest Europe. *In* Kay, M. (Ed.) North Atlantic—Geology and continental drift: a symposium. *Mem. Am. Ass. Petrol. Geol.* **12**, 557–62.

—— 1971a. The age and correlation of the Rosroe Grits, north-west Co. Galway. *Proc. R. Ir. Acad.* Sec. B. **71**, 75–83.

—— 1971b. Palaeontological evidence bearing on the age of Dalradian deformation and metamorphism in Ireland and Scotland. *Scott. J. Geol.* **7**, 285–8.

—— 1972. *Monograptus priodon* (Bronn) from the Macduff Group (Upper Dalradian) of Banffshire, Scotland. *Geol. Mag.* **108**, 485–7.

SMITH, B. 1912. The glaciation of the Black Combe district. *Q. Jl geol. Soc. Lond.* **68**, 402–48.

—— 1924. The unconformable base of the Coniston Limestone Series in the Lake District. *Geol. Mag.* **61**, 163–7.

—— 1935. The Mynydd Cricor Inlier. *Proc. Geol. Ass.* **46**, 187–92. [52].

SMITH, R. A. 1965. *A bibliography of Lake District geology and geomorphology.* Cumberland Geological Society, Whitehaven. 1–44.

SMITH, S. 1933. On the occurrence of Tremadoc shales in the Tortworth Inlier (Gloucestershire). With notes on the fossils by Cyril James Stubblefield. *Q. Jl geol. Soc. Lond.* **89**, 357–78.

SOPER, N. J. 1970. Three critical localities on the junction of the Borrowdale volcanic rocks with the Skiddaw Slates in the Lake District. *Proc. Yorks. geol. Soc.* **37**, 461–93.

SORBY, H. C. 1908. On the application of quantitative methods to the study of the structure and history of rocks. *Q. Jl geol. Soc. Lond.* **64**, 171–233.

SPJELDNAES, N. 1957. The Middle Ordovician of the Oslo region, Norway. 8. Brachiopods of the Suborder Strophomenida. *Norsk geol. Tidsskr.* **37**, 1–214.

—— 1963. Some silicified Ordovician fossils from South Wales. *Palaeontology* **6**, 254–63.

STAMP, L. D. & WOOLDRIDGE, S. W. 1923. The igneous and associated rocks of Llanwrtyd (Brecon). *Q. Jl geol. Soc. Lond.* **79**, 16–45. [91].

STANTON, W. I. 1960. The Lower Palaeozoic rocks of south-west Murrisk, Ireland. *Q. Jl geol. Soc. Lond.* **114**, 269–96.

STEVENSON, I. P. 1971. The Ordovician rocks of the country between Dwygyfylchi and Dolgarrog, Caernarvonshire. *Proc. Yorks. geol. Soc.* **38**, 517–48. [34].

STRAHAN, A., CANTRILL, T. C., DIXON, E. E. L. & THOMAS, H. H. 1907. The geology of the South Wales Coalfield. Part VII. The country around Ammanford. (Sheet 230). *Mem. geol. Surv. U.K.*, i–viii, 1–246. [112].

——, ——, —— & —— 1909. The geology of the South Wales Coalfield. Part X. The country around Carmarthen. (Sheet 229). *Mem. geol. Surv. U.K.*, i–viii, 1–177. [111].

——, ——, ——, —— & JONES, O. T. 1914. The geology of the South Wales Coalfield. Part XI. The country around Haverfordwest. (Sheet 228). *Mem. geol. Surv. U.K.*, i–viii, 1–262. [110].

STUBBLEFIELD, C. J. 1967. Some results of a recent Geological Survey boring in Huntingdonshire. *Proc. geol. Soc. Lond.* **1637**, 35–40.

—— & BULMAN, O. M. B. 1927. The Shineton Shales of the Wrekin district: with notes on their development in other parts of Shropshire and Herefordshire. *Q. Jl geol. Soc. Lond.* **83**, 96–146. [71].

SWANSTON, W. 1876–77. On the Silurian rocks of the County Down. *Rep. Proc. Belf. Nat. Fld Club* Ser. 2. **1**, appendix IV, 107–23.

SWEET, W. C. & BERGSTRÖM, S. M. 1971. The American Upper Ordovician Standard XIII: a revised time-stratigraphic classification of North American upper Middle and Upper Ordovician rocks. *Bull. geol. Soc. Am.* **82**, 613–28.

SWETT, K. 1969. Interpretation of depositional and diagenetic history of Cambrian-Ordovician succession of Northwest Scotland. *In* Kay, M. (Ed.) North Atlantic—geology and continental drift: a symposium. *Mem. Am. Ass. Petrol. Geol.* **12**, 630–46.

TEMPLE, J. T. 1952. A revision of the trilobite *Dalmanitina mucronata* (Brongniart) and related species. *Acta Univ. lund. N.F. Avd. 2.* **48**, 1–33.

—— 1968. The Lower Llandovery (Silurian) brachiopods from Keisley, Westmorland. *Palaeontogr. Soc. (Monogr.)*, 1–58.

THEOKRITOFF, G. 1951. Ordovician rocks near Leenane, Ireland. *Proc. R. Ir. Acad. Sec. B.* **54**, 25–49.

THOMAS, D. E. 1960. The zonal distribution of Australian graptolites. *J. Proc. R. Soc. N.S.W.* **94**, 1–58.

THOMAS, G. E. & THOMAS, T. M. 1956. The volcanic rocks of the area between Fishguard and Strumble Head, Pembrokeshire. *Q. Jl geol. Soc. Lond.* **112**, 291–314. [101].

THOMAS, H. H. 1911. The Skomer Volcanic Series (Pembrokeshire). *Q. Jl geol. Soc. Lond.* **67**, 175–214.

—— & COX, A. H. 1924. The Volcanic Series of Trefgarn, Roch and Ambleston (Pembrokeshire). *Q. Jl geol. Soc. Lond.* **80**, 520–48. [105].

—— & JONES, O. T. 1912. On the pre-Cambrian and Cambrian rocks of Brawdy, Hayscastle and Brimaston (Pembrokeshire). *Q. Jl geol. Soc. Lond.* **68**, 374–401. [106].

TOGHILL, P. 1970a. A fauna from the Hendre Shales (Llandeilo) of the Mydrim area, Carmarthenshire. *Proc. geol. Soc. Lond.* **1663**, 121–9.

—— 1970b. Highest Ordovician (Hartfell Shales) graptolite faunas from the Moffat area, south Scotland. *Bull. Br. Mus. nat. Hist. (Geol.)* **19**, 1–26.

TREMLETT, W. E. 1959. The structure of the Lower Palaeozoic rocks of the Arklow district (Ireland). *Q. Jl geol. Soc. Lond.* **115**, 17–40.

—— 1962. The geology of the Nefyn-Llanaelhaiarn area of North Wales. *Lpool Manchr geol. J.* **3**, 157–76. [9].

—— 1964. The geology of the Clynnog-fawr district and Gurn Ddu Hills of northeast Lleyn. *Geol. J.* **4**, 207–23. [11].

—— 1965. The geology of the Chwilog area of southeastern Lleyn (Caernarvonshire). *Geol. J.* **4,** 435–48. [10].

—— 1969. Caradocian volcanicity in the Lleyn peninsula. *In* Wood, A. (Ed.) *The Pre-Cambrian and Lower Palaeozoic rocks of Wales.* University of Wales Press, Cardiff, 357–85.

TWENHOFEL, W. K. *et al.* 1954. Correlation of the Ordovician formations of North America. With an Introduction by Dunbar, C. O., and a Discussion of British-American correlation by Whittington, H. B. *Bull. geol. Soc. Am.* **65,** 247–98.

TURNER, J. S. 1961. [Report of a field meeting to Cautley]. *Proc. Yorks. geol. Soc.* **33,** 36–7.

ULRICH, E. O., FOERSTE, A. F. & MILLER, A. K. 1943. Ozarkian and Canadian cephalopods, Part 2: Brevicones. *Spec. Pap. geol. Soc. Am.* **49,** i–x, 1–240.

——, ——, —— & UNKLESBAY, A. G. 1944. Ozarkian and Canadian cephalopods, Part 3: Longicones and summary. *Spec. Pap. geol. Soc. Am.* **58,** i–x, 1–226.

WADE, A. 1911. The Llandovery and associated rocks of north-eastern Montgomeryshire. *Q. Jl geol. Soc. Lond.* **67,** 415–59. [65].

WADGE, A. J. 1972. Sections through the Skiddaw-Borrowdale unconformity in eastern Lakeland. *Proc. Yorks. geol. Soc.* **39** (in press).

——, NUTT, M. J. C., LISTER, T. R. & SKEVINGTON, D. 1970. A *Didymograptus murchisoni* Zone fauna from the Lake District. *Geol. Mag.* **106,** 595–8.

WALTHAM, A. C. 1971. A note on the structure and succession at Abereiddy Bay, Pembrokeshire. *Geol. Mag.* **108,** 49–52. [103].

WARD, J. C. 1876. The geology of the northern part of the English Lake District. *Mem. geol. Surv. U.K.,* i–xii, 1–132.

WATTS, W. W. 1885. On the igneous and associated rocks of the Breidden Hills in east Montgomeryshire and west Shropshire. *Q. Jl geol. Soc. Lond.* **41,** 532–46.

—— 1925. The geology of south Shropshire. *Proc. Geol. Ass.* **36,** 321–63.

WEDD, C. B. 1932. Notes on the Ordovician rocks of Bausley, Montgomeryshire. *Summ. Progr. geol. Surv. Lond.* (for 1931), Pt. 2, 49–55.

——, SMITH, B., KING, W. B. R. & WRAY, D. A. 1929. The geology of the country around Oswestry. (Sheet 137). *Mem. geol. Surv. U.K.,* i–xix, 1–234. [61].

——, —— & WILLS, L. J. 1927. The geology of the country around Wrexham, Part 1. Lower Palaeozoic & Lower Carboniferous rocks. (Sheet 121). *Mem. geol. Surv. U.K.,* i–xviii, 1–179. [55].

WEIR, J. A. 1962. Geology of the Lower Palaeozoic inliers of Slieve Bernagh and the Cratloe Hills, County Clare. *Scient. Proc. R. Dubl. Soc.* Ser. A. **1,** 233–63.

WELLS, A. K. 1925. The geology of the Rhobell Fawr district (Merioneth). *Q. Jl geol. Soc. Lond.* **81,** 463–538. [38].

WHITAKER, W. & JUKES-BROWNE, A. J. 1894. On deep borings at Culford and Winkfield, with notes on those at Ware and Cheshunt. *Q. Jl geol. Soc. Lond.* **50,** 488–514.

WHITTARD, W. F. 1931. The geology of the Ordovician and Valentian rocks of the Shelve country, Shropshire. *Proc. Geol. Ass.* **42,** 322–39.

—— 1932. The stratigraphy of the Valentian rocks of Shropshire. The Longmynd-Shelve and Breidden outcrops. *Q. Jl geol. Soc. Lond.* **88,** 859–902. [66].

—— 1952. A geology of south Shropshire. *Proc. Geol. Ass.* **63,** 143–97.

—— 1955–1967. The Ordovician trilobites of the Shelve Inlier, west Shropshire. *Palaeontogr. Soc. (Monogr.)* Pt. I, 1–40, 1955; Pt. II, 41–70, 1956; Pt. III (for 1957), 71–116, 1958; Pt. IV (for 1959), 117–62, 1960; Pt. V (for 1960), 163–96, 1961; Pt. VI, 197–228, 1961; Pt. VII (for 1963), 229–64, 1964; Pt. VIII (for 1965), 265–306, 1966; Pt. IX, 307–52, 1967.

—— (Ed.) 1960. *Lexique Stratigraphique International. Congrès Géologique International—Commission de Stratigraphie.* Volume I. Europe. Fascicule 3a England, Wales & Scotland, Part 3a IV Ordovician. Centre National de la Recherche Scientifique, Paris, 1–296.

WHITTINGTON, H. B. 1938. The geology of the district around Llansantffraid ym Mechain, Montgomeryshire. *Q. Jl geol. Soc. Lond.* **94,** 423–57. [60].

—— 1950. Sixteen Ordovician genotype trilobites. *J. Paleont.* **24,** 531–65.

—— 1952. The trilobite family Dionididae. *J. Paleont.* **26,** 1–11.

—— 1962–1968. The Ordovician trilobites of the Bala area, Merioneth. *Palaeontogr. Soc.* (*Monogr.*) Pt. I, 1–32, 1962; Pt. II (for 1964), 33–62, 1965; Pt. III, 63–92, 1966; Pt. IV, 93–138, 1968.

—— 1966. Trilobites of the Henllan Ash, Arenig Series, Merioneth. *Bull. Br. Mus. nat. Hist.* (Geol.) **11**, 489–505.

—— 1968. Zonation and correlation of Canadian and Early Mohawkian Series. *In* Zen, E-an, White, W. S., Hadley, J. B. & Thompson, J. B. (Eds.) *Studies of Appalachian geology: northern and maritime.* Interscience, Wiley, New York, 49–60.

—— & HUGHES, C. P. 1972. Ordovician geography and faunal provinces deduced from trilobite distribution. *Phil. Trans. R. Soc.* Ser. B. **263**, 235–78.

—— & WILLIAMS, A. 1955. The fauna of the Derfel Limestone of the Arenig district, North Wales. *Phil. Trans. R. Soc.* Ser. B. **238**, 397–427.

—— & —— 1961. The Ordovician period. *In* Harland, W. B., Smith, A. G. & Wilcock, B. (Eds.) The Phanerozoic Time-scale. *Q. Jl geol. Soc. Lond.* 120s, 241–54.

WILLIAMS, A. 1948. The Lower Ordovician cryptolithids of the Llandeilo district. *Geol. Mag.* **85**, 65–88.

—— 1949. New Lower Ordovician brachiopods from the Llandeilo-Llangadock district. Parts I and II. *Geol. Mag.* **86**, 161–74, 226–38.

—— 1951. Llandovery brachiopods from Wales with special reference to the Llandovery district. *Q. Jl geol. Soc. Lond.* **107**, 85–134.

—— 1953. The geology of the Llandeilo district, Carmarthenshire. *Q. Jl geol. Soc. Lond.* **108**, 177–208. [87].

—— 1956. *Productorthis* in Ireland. *Proc. R. Ir. Acad.* Sec. B. **57**, 179–83.

—— 1962. The Barr and Lower Ardmillan Series (Caradoc) of the Girvan district, south-west Ayrshire, with descriptions of the Brachiopoda. *Mem. geol. Soc. Lond.* **3**, 1–267.

—— 1963. The Caradocian brachiopod faunas of the Bala district, Merionethshire. *Bull. Br. Mus. nat. Hist.* (Geol.) **8**, 327–471.

—— 1969a. Ordovician faunal provinces with reference to brachiopod distribution. *In* Wood, A. (Ed.) *The Pre-Cambrian and Lower Palaeozoic rocks of Wales.* University of Wales Press, Cardiff, 117–54.

—— 1969b. Ordovician of the British Isles. *In* Kay, M. (Ed.) North Atlantic—geology and continental drift: a symposium. *Mem. Am. Ass. Petrol. Geol.* **12**, 236–64.

—— 1972a. An Ordovician Whiterock fauna in western Ireland. *Proc. R. Ir. Acad.* Sec. B. **72**, 209–19.

—— 1972b. Distribution of brachiopod assemblages in relation to Ordovician palaeogeography. *Spec. pap. palaeont. Ass.* (in press).

WILLIAMS, D. 1930. The geology of the country between Nant Peris and Nant Ffrancon (Snowdonia). *Q. Jl geol. Soc. Lond.* **86**, 191–233. [28].

WILLIAMS, H. 1922. The igneous rocks of the Capel Curig district (North Wales). *Proc. Lpool geol. Soc.* **13**, 166–206. [31].

—— 1927. The geology of Snowdon (North Wales). *Q. Jl geol. Soc. Lond.* **83**, 346–431. [21].

—— & BULMAN, O. M. B. 1931. The geology of the Dolwyddelan syncline (North Wales). *Q. Jl geol. Soc. Lond.* **87**, 425–58. [30].

WILLIAMS, T. G. 1934. The Pre-Cambrian and Lower Palaeozoic rocks of the eastern end of the St. David's Pre-Cambrian area, Pembrokeshire. *Q. Jl geol. Soc. Lond.* **90**, 32–75. [104].

WILLS, L. J. & SMITH, B. 1922. The Lower Palaeozoic rocks of the Llangollen district with special reference to the tectonics. *Q. Jl geol. Soc. Lond.* **78**, 176–226. [53].

WOOD, D. S. 1969. The base and correlation of the Cambrian rocks of North Wales. *In* Wood, A. (Ed.) *The Pre-Cambrian and Lower Palaeozoic rocks of Wales.* University of Wales Press, Cardiff, 47–66.

—— & HARPER, J. C. 1962. Notes on a temporary section in the Ordovician at Conway, North Wales. *Lpool Manchr geol. J.* **3**, 177–86.

WOOLACOTT, D. 1923. On a boring at Roddymoor Colliery, near Crook, Co. Durham. *Geol. Mag.* **60**, 50–62.

WRIGHT, A. D. 1963. The fauna of the Portrane Limestone. I. The inarticulate brachiopods. *Bull. Br. Mus. nat. Hist.* (Geol.) **8**, 221–54.

—— 1964. The fauna of the Portrane Limestone, II. *Bull. Br. Mus. nat. Hist.* (Geol.) **9**, 157–256.

—— 1967. A note on the stratigraphy of the Kildare Inlier. *Ir. Nat. J.* **15**, 340–3.

—— 1968. A westward extension of the Upper Ashgillian *Hirnantia* fauna. *Lethaia* **1**, 352–67.

—— 1970. The stratigraphic distribution of the Ordovician inarticulate brachiopod *Orthisocrania divaricata* (M'Coy) in the British Isles. *Geol. Mag.* **107**, 97–103.

WRIGHT, J. E. 1968. *The geology of the Church Stretton area. (Explanation of 1:25 000 Geological Sheet SO49).* Institute of Geological Sciences. H.M.S.O., London, i–vi, 1–89. [68].

YOCHELSON, E. L. 1964. The early Ordovician gastropod *Ceratopea* from east Greenland. *Meddr. Grønland* **164**, 1–12.

ZEN, E-AN, WHITE, W. S., HADLEY, J. B. & THOMPSON, J. B. (Eds.) 1968. *Studies of Appalachian geology: northern and maritime.* Interscience, Wiley, New York, i–xviii, 1–475.

ZIEGLER, A. M., McKERROW, W. S., BURNE, R. V. & BAKER, P. E. 1969. Correlation and environmental setting of the Skomer Volcanic Group, Pembrokeshire. *Proc. Geol. Ass.* **80**, 409–39.

WILLIAMS, PROF. A., PH. D., F.R.S., F.R.S.E., M.R.I.A., F.G.S.
Department of Geology, The Queen's University, Belfast.

STRACHAN, I., PH. D., F.G.S.
Department of Geology, The University, Birmingham.

BASSETT, D. A., PH. D., F.G.S.
Department of Geology, National Museum of Wales, Cardiff.

DEAN, W. T., PH. D., F.G.S.
Geological Survey of Canada, Ottawa.

INGHAM, J. K., PH. D., F.G.S..
Hunterian Museum, Department of Geology, The University, Glasgow.

WRIGHT, A. D., PH. D., F.G.S.
Department of Geology, The Queen's University, Belfast.

WHITTINGTON, PROF. H. B., D. SC., PH. D., A.M., F.R.S., F.G.S.
Sedgwick Museum, Department of Geology, The University, Cambridge.

Much appreciated grants in aid of publication have been received from the National Museum of Wales and The Queen's University of Belfast.

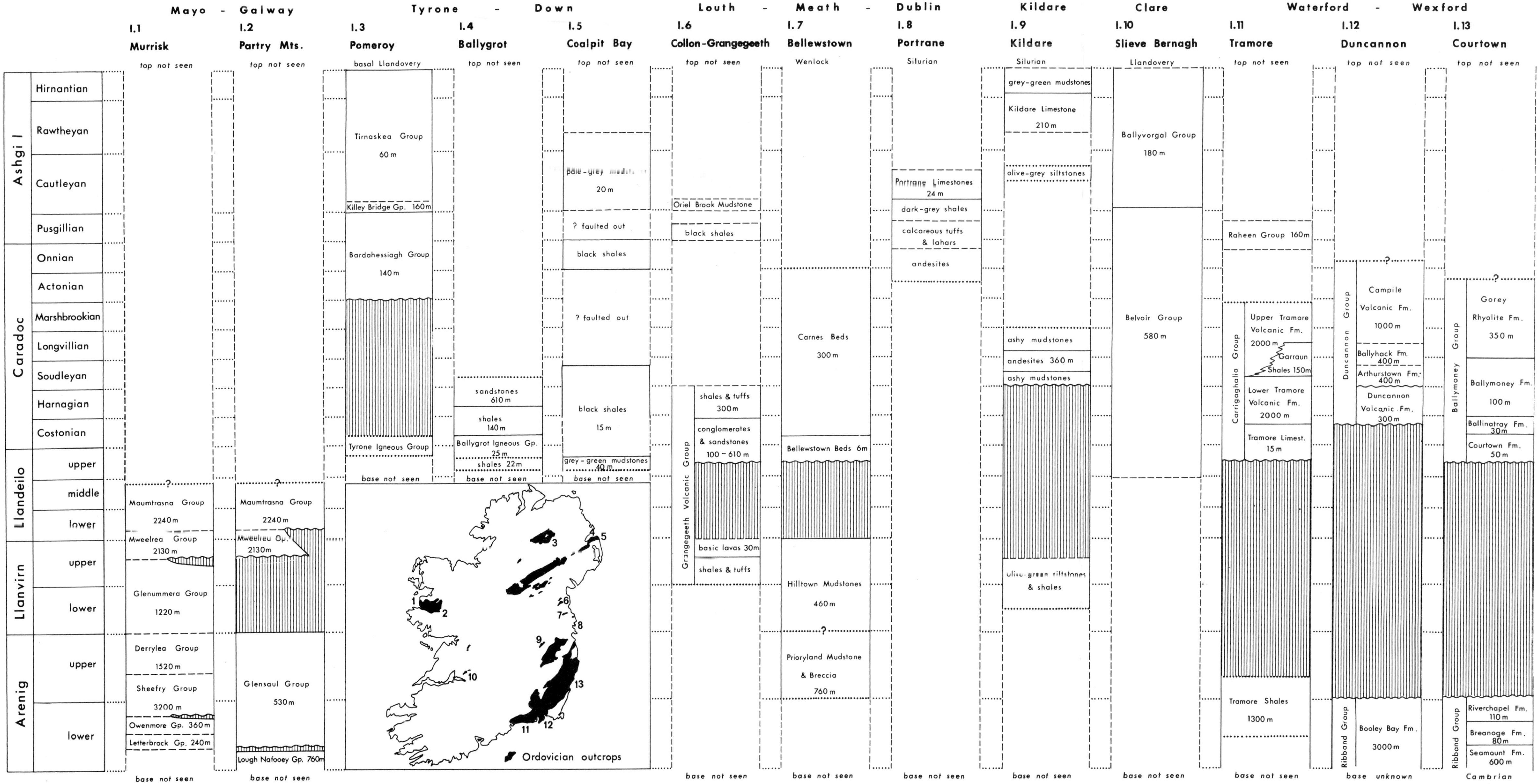

FIG. 10—Chart showing a correlation of selected Ordovician successions exposed in Ireland.